Equitable Access to Human Biological Resources in Developing Countries

Roger Scarlin Chennells

Equitable Access to Human Biological Resources in Developing Countries

Benefit Sharing Without Undue Inducement

Springer

Roger Scarlin Chennells
Chennells Albertyn Attorneys
Stellenbosch, Cape Town
South Africa

ISBN 978-3-319-36974-7 ISBN 978-3-319-19725-8 (eBook)
DOI 10.1007/978-3-319-19725-8

Springer Cham Heidelberg New York Dordrecht London

Softcover reprint of the hardcover 1st edition 2016

Printed on acid-free paper

Springer International Publishing AG Switzerland is part of Springer Science+Business Media
(www.springer.com)

I would like to dedicate this book
To Jono, Sparks, Bids, Miah and Jilly.

Endorsements

Indigenous peoples' genes are precious resources for research and innovation. Yet, no binding international law governs access and benefit sharing for human DNA. In this in-depth ethical and legal study, Chennells makes a convincing case for benefit sharing. With his engaging style, he succeeds in bringing a complex area to life and provides valuable advice for indigenous peoples, policy makers, ethics committees, academics and NGOs. Strongly recommended reading.

Prof. Doris Schroeder
Director of the Centre for Professional Ethics, UCLAN, UK

Genomic studies to date have shown quite convincingly that the San from Southern Africa have retained some of the oldest genetic signatures found among living people throughout the world. Consequently, their genetic material is much sought after by researchers conducting genomic studies to understand the molecular evolutionary processes that have contributed in generating the genetic diversity seen among living people today. How do we balance scientific curiosity for knowledge with social issues related to good ethical practice, respect for research subjects, and acknowledgment of the plight of indigenous communities, engaging with, and educating the public around science? Roger Chennells has worked with the San and represented them on legal issues in the past. In this book, he brings the humanitarian issues into focus, reminds us about governance as stipulated in the legal framework, but most of all, raises our collective consciousness and awareness to sensitive issues, which when addressed respectfully through engagement, has the potential for win–win outcomes for all.

Prof. Himla Soodyall
Division of Human Genetics
National Health Laboratory Service
and University of the Witwatersrand
Johannesburg, South Africa

Since the Convention on Biological Diversity was adopted in 1992, much has been written on access to genetic resources and benefit sharing, and the international rules are now in place. Human biodiversity is another source of genetic material of potential value, but we have no clear guidelines on how to regulate fair and culturally appropriate exchange transactions involving indigenous peoples, government, and industry. This excellent volume forms a practical and philosophical basis to meet the ethical, legal, and regulatory challenges, benefiting from the author's unique insights as a legal representative of indigenous groups in Southern Africa.

Prof. Graham Dutfield
Professor of International Governance
University of Leeds, UK

Acknowledgments

This book is based upon my Ph.D. thesis during which I was ably supported by my supervisors Doris Schroeder and Peter Herissone-Kelly, as well as funded by a bursary from the Wellcome Trust.

I wish to also acknowledge and single out certain San leaders with whom I have worked for many years as their lawyer, and who have been a source of wisdom, inspiration, occasional despair and regular humour. They are Andries Steenkamp, Mario Mahongo, Collin Louw, Zeka Shiwarra, Leana Snyders and Mathambo Ngakaeaja. Their individual and collective struggles over decades to advance the rights of the San provided me with concrete reasons to write this book.

Finally, my children Rebecca, Guy, Oliver, Clara and Sebastian were often bemused but always supportive during this project, Bex pouring encouraging cups of Earl Grey tea, and David reminding me that 'retreat is not an option.' Biddy and Andrew provided generous home comforts and a safe corner in London, and my partner Judy Beaumont's loyal motivation and support contributed hugely towards the eventual completion.

Abbreviations

ACUNS	Association of Canadian Universities on Northern Studies
AIATSIS	Australian Institute of Aboriginal and Torres Straits Islander Studies
CBD	Convention on Biological Diversity
COHRED	Council on Health Research for Development
DNA	Deoxyribonucleic Acid
EC	Ethics Committee
EPO	European Patent Office
GWAS	Genome-Wide Association Studies
HGDP	Human Genome Diversity Project
HPGR	Human Population Genetic Research
HUGO	Human Genome Organisation
ICESCR	International Covenant on Economic, Social and Cultural Rights
IHS	Indian Health Service
IPACC	Indigenous Peoples of Africa Coordinating Committee
IPCB	Indigenous Peoples Council on Biocolonialism
IPHRC	Indigenous Peoples' Health Research Centre
IRB	Institutional Review Board
IWRI	Indigenous Wellness Research Institute in America
LMIC	Low Middle Income Country
NBAC	National Bioethics Advisory Commission
NHGRI	National Human Genome Research Institute
NHMRC	National Health and Medical Research Council
NIH	National Institutes of Health
REC	Research Ethics Committee
SPEAR	Social Policy Evaluation and Research Committee
TRIPS	Trade Related Aspects of Intellectual Property Rights
UNEP	United Nations Environment Programme
UNESCO	United Nations Educational, Scientific and Cultural Organisation
UNOOSA	United Nations Office for Outer Space Affairs
UNPFII	United Nations Permanent Forum on Indigenous Issues
USPTO	United States Patent and Trade Office

WHO	World Health Organisation
WIPO	World Intellectual Property Organisation
WMA	World Medical Association
WIMSA	Working Group of Indigenous Minorities in South Africa

List of Tables

Abstract

The overall question examined in this book is, how can cross-border access to human genetic resources, such as blood or DNA samples, be governed to achieve equity for developing countries?

Access to and benefit sharing for human biological resources is not regulated through an international legal framework such as the Convention on Biological Diversity, which applies only to plants, animals and microorganisms as well as associated traditional knowledge. This legal vacuum for the governance of human genetic resources can be attributed (in part) to the concern that benefit sharing might provide undue inducements to research participants and their communities.

This book aims to show that:

(a) Benefit sharing is crucial to avoiding the exploitation of developing countries in genetic research.
(b) With functioning research ethics committees, undue inducement is less of a concern in genetic research than in other medical research (e.g. clinical trials).
(c) Justified concerns remain regarding genetic research involving indigenous populations. These concerns are debated, and some recommendations are provided.

In addressing the above highly pressing topics in global bioethics and international law, the book combines bioethical argument with jurisprudence, drawing on the law of equity as well as the legal concepts of duress (coercion), unconscionable dealing, and undue influence. Recommendations are made towards more ethical genetic research involving indigenous peoples.

Chapter 1
Introduction

Abstract The introductory chapter places the topic of exchange of human biological resources within the parameters of modern human population genetic research (HPGR). In doing so it traverses the legal and ethical landscape pertaining to population wide genomic research, including population bio-banks and global bio-medical research, describing the types of cases that have aroused persistent perceptions of unfairness. These cases lead to brief discussions on the issues raised in more depth during the book. In particular the topic of exploitation encountered during genetic and genomic research is introduced, and institutional legal recommendations for benefit sharing are described in the context of theories in bioethics that oppose benefit sharing in principle. Finally ethical concerns about the dangers of undue inducement are described, as well as the specific problems pertaining to genomic research which is conducted on vulnerable and in particular indigenous research populations.

Keywords Human biological resources · Human population genetic research · Research ethics · Benefit sharing · Indigenous research populations

Humans have since ancient times sought to unravel the mysteries of life. This quest, involving fields of knowledge ranging from life science to philosophy, has a mysterious and even existential component, delving as it does into the very heart of the human condition. Human DNA has been at the centre of relevant research endeavours since the 20th Century. This book examines the equity or fairness of the exchange of human biological resources, in particular DNA samples, across national borders. This exchange is very important for scientific and health research; yet it has been described as highly exploitative of developing countries (Macklin 2004, p. 68).[1] How to enable scientific and medical research using

[1]The commonly used descriptive adjectives, *developed* and *developing* countries are utilised throughout this book rather than, 'first and third world', 'north and south', or other formulations currently in use.

R.S. Chennells, *Equitable Access to Human Biological Resources in Developing Countries*, DOI 10.1007/978-3-319-19725-8_1

human biological resources without exploiting individuals, and in particular vulnerable populations such as the indigenous or Aboriginal peoples frequently targeted for genetic research, is the main question of this book.

In order to resolve this question, a number of sub-topics have to be examined, in particular:

1. How does the putative exploitation of donors of human biological resources manifest itself at the beginning of the 21st Century?
2. What is the legal and ethical landscape governing this area of enquiry? That is are there existing legal or ethical frameworks, which aim to protect relevant groups from exploitation? If so, what are their deficits given on-going exploitation claims? And how can these be addressed?
3. Working towards answering these questions requires clarity on the concepts used. In particular it needs to be clear what I mean by genetic and genomic research, and by human biological resources. In addition, other terms such as benefit sharing, exploitation, equity, common heritage of humankind, altruism, solidarity, indigenous peoples, international customary law, commodification, and undue inducement will become important as the book progresses and will be explained as they arise.

At the same time a number of questions fall outside the scope of this investigation. For example, questions about the legitimacy of genomic research. To illustrate this point, one might argue that the public funds spent on genomic research on malaria could be better used to distribute bed-nets.[2] However, I shall assume with Gro Harlem Brundtlandt the former Director General of the World Health Organisation (WHO 2002) that, "it is clear that the science of genomics holds tremendous potential for improving health globally [...] the specific challenge is how to harness this knowledge and have it contribute towards health equity, especially amongst developing nations" (ibid, p. 3).

Other important questions such as the potential exploitation of clinical trial participants or researchers, of forensic databases created by governments for purposes of resolving crime and enforcing law and order also fall outside the scope of this book. Broader questions about the human right to access to health care as well as major issues in global distributive justice (Pogge 2008) cannot be ignored and will be used for context, but will also not be addressed in full depth.

Finally and with moral emphasis I intend to address the question of the fair and equitable treatment of indigenous peoples in genomic research, and to provide convincing arguments to persuade or counter the beliefs of an imaginary sceptic whose views, a composite of various authors who will be referred to below, could be condensed and expressed as follows,

---

[2]Question from the floor at the Human Heredity and Health (H3) Conference, Cape Town 3–6 March 2011.

> Exploitation is not a problem in genetic research, for indigenous peoples or any other communities. Benefit sharing is inappropriate for a number of reasons, including the fact that human genetic resources are part of the common heritage of humankind. In addition, individuals should share willingly out of altruism, and they have done nothing to add value to their DNA. If benefit sharing were to be allowed, (which we do not concede) it would run the risk of unduly influencing or coercing poor research participants. Finally, indigenous peoples are but one of many types of genomic research communities, and their concerns are adequately addressed by mainstream research ethics guidelines.

In order to achieve the objective set out above and to counter the views of the imaginary sceptic, the origin of the notion of benefit sharing will need to be broadly examined in this introduction, this mechanism having been proposed in order to prevent exploitation during the exchange of genetic materials for research. Three leading questions will be specifically addressed in the book, namely firstly whether and in what circumstances benefit sharing is appropriate for human genetic research, secondly whether such benefit sharing runs the risk of unduly inducing or coercing research participants, and thirdly whether the particular concerns of indigenous peoples are adequately met by the current research practice and guidelines.

For the remainder of this introductory chapter one shall describe the field of genetic and genomic research involving human biological resources, proceeding to a description of global research today.[3] Problems occurring in the exchange of human genetic samples for research purposes will then be described, with reference to a number of cases that have been viewed as exploitative or unfair in various ways, which form the backdrop to and motivation of the research questions. The chosen cases involve communities described as indigenous peoples, whose concerns about and responses to these issues will be examined as a priority in this book. Apprehensions about the fairness of genomic research encounters with such communities in the developing world will then be examined, leading to a description of proposals for benefit sharing to be introduced as a mechanism to counter the potential unfairness in the exchange of genetic material for research. The chapter will close with an outline of how the research questions will be addressed during the remainder of the book.

[3]The term genomic research is used more often in this book for the reason that it is broader in scope and includes genetic research. In some contexts the words are used as being more or less interchangeable.

1.1 Genetics, Genomics and Human Biological Resources

In 1794, Erasmus Darwin, grandfather of the more famous Charles, a renowned polymath poet and physician of his time, predicted the discovery of a single basis of all organic life, one that was later to be termed as DNA.[4]

> As the earth and the ocean were probably peopled with vegetable productions long before the existence of animals: and many families of these animals long before other families of them, shall we conjecture that one and the same kind of living filaments is and has been the cause of all organic life? (Darwin 1794, p. 244)

Exponential advances in biotechnology over the past decades have exposed the essential building blocks of nature, termed by some, "God's blueprints for life" (Ridley 1999, p. 8), in their fascinating complexity. Popular authors such as Dawkins (1995) and Matt Ridley have contributed towards placing the idea firmly in the public domain, and human genetic resources have become a new and valuable commodity, their exchange fuelled by and contained within a rapidly transmuting intellectual property rights system.

Exploration of the human genome has proceeded apace, triggered by the discovery of the structure of DNA by Watson and Crick in 1953, and characterized by a dynamism and breadth that has transformed the paradigms previously underpinning science and public health. Watson and Crick's discovery of the DNA double helix can be equated with Darwin's theory of natural selection as a milestone in human history in its presaging of a new era of biological and scientific exploration. The heightened importance of human genetic resources over the past decades has contributed towards an explosion in genomic research, has forced states to revisit their laws, and has given rise to fresh ethical and legal challenges.

The term *human biological resources* refers to all and any parts of the body including tissue, bone and fluids, whereas *human genetic resources* is a more narrow term denoting the components of human DNA. The latter comprises the raw material utilised by the vast biomedical research industry, the appropriate utilisation of which is the intended subject of this book. A brief description of human DNA commences with cells, which exist as the basic working units of every living system, and with the chemical deoxyribonucleic acid (DNA) within each such cell which contains the full instructions to create that entire individual (Mehlman and Botkin 1998). These instructions are located within the chromosomes, namely twenty-three pairs of individual DNA strands, which in humans contain an

[4]According to the Joy of Knowledge Encyclopedia DNA or deoxyribononucleic acid is a long strand of matter too fine to be seen with the most powerful optical microscope, arranged in the form of a twisted rope ladder with millions of rungs—the double helix. The struts are made up of alternating units of phosphate and deoxyribose sugar. Each rung contains a linked pair of chemical compounds called nucleic acid bases.

estimated thirty thousand encoded genes (The International Human Genome Sequencing Consortium).[5] Genes provide codes for proteins, which in turn determine the structure and characteristics of all life forms (Ridley 1999, pp. 6–9).

Genes used to be regarded in abstract form as the unit of inheritance that transferred characteristics from parent to child, but have now with the advancement of molecular biology become physically observable units, sequences of DNA which when converted into strands of what is called messenger RNA become the basis for building the associated proteins. As Richard Dawkins explains in his book *The Selfish Gene*, in which he describes the central role played by genes in evolution, "genes are denizens of geological time: genes are forever." This message was recorded more playfully in the following verse form

> An itinerant selfish gene
>
> Said, 'bodies aplenty I've seen.
>
> You think you're so clever,
>
> But I'll live forever,
>
> You're just a survival machine (Dawkins 2004, p. 630).

Genetics, the study of heredity in general, and genes in particular (Jones 2009, p. 3) is an exploding field, described by some as not only the science of the 21st Century but also and in equal measure as big business, with applications from medical diagnosis to drug production, law enforcement to chicken breeding (Jones 2009, p. ix). The genome contains the sum total of genetic information of an individual, which information resides in each cell of the body and is encoded in the structure of the DNA (ibid, p. 28).[6] The DNA sequence of each organism consists of the particular arrangement of bases along the DNA strand, and the science of genomics, as described by one commentator, now allows the *metaphor* of life-as-information to become *material* reality, that can be commodified (Rajan 2006, p. 16). How this specific form of information is utilised by research for eventual commodification is discussed in more detail below.

Genetic information, which relates to families as well as to individuals, is the research target of all genetic research, and has been differentiated by Zhou into three different types. Firstly, non-differential human genetic information, secondly, individual human genetic information, and thirdly, group/collective human genetic information (Zhou 2006, p. 113). Non-differential human genetic information is the genetic information shared by each human being, namely the approximately

---

[5]This number has been estimated as being closer to 20,000 genes by the National Human Genome Research institute. [online] Available at http://www.genome.gov/DNADay/q.cfm?aid=2&year=2012.

[6]The Human genome, like the genome of all other living animals, is a collection of long polymers of DNA. These polymers are maintained in duplicate copy in the form of chromosomes in every human cell and encode in their sequence of constituent bases (G, A, T and C) the details of the molecular and physical characteristics that form the corresponding organism. The sequence of these polymers, their organisation and structure, provide the genome with the capability to replicate, repair, package and maintain itself.

99.8 % of the information on the human's approximately 30,000 genes that is entirely uniform. Neither states nor individuals could claim any proprietary rights over this wealth of shared information, which is potentially of great benefit to humankind, and could aptly fall into the category of common heritage of humankind, which will be discussed later. Individual human genetic information refers to the 0.02 %, which determines the diversity of the human race, from outward appearance to susceptibility to diseases, and is unique to each human. Bio-banks collect this type of genetic information, and it is highly prized, being the particular focus of forensic databases, relationship testing and genetic research on diseases (Richards 2001, p. 674). According to Annas, this form of personal information is, in some countries at least, the subject of a right to privacy (Annas 1999, p. 7).

With regards to community or group/collective perspectives human genetic information is a combination of the first two types described above, namely non-differential and individual human genetic information, but in addition concerns the genetic properties of groups. As McGregor explains, a group of people living together for a sufficiently long period of time are likely to share genetic properties, which become a research tool to explore variations that can lead to knowledge about genetic disorders, and the origins and migration patterns of peoples (McGregor 2007). These communities can become dispersed, such as the Ashkenazi Jews who share genetic mutations predisposing them to certain diseases, but nevertheless retain their collective association as a genetic group (Weijer 1999, p. 502). This form of genetic information is the research target of human population genetic research (HPGR) and is arguably the subject of certain group as opposed to individual rights (Wang 2011, p. 10).

Genetic research examines how genes relate to environmental factors and the health of humans, whilst genomic research studies focus on the whole genome, usually across selected populations (Gibbons et al. 2007). As might be anticipated, the rapid expansion of the interrelated fields described below has exposed the need for regulatory systems relating to research use of human tissue to adapt to changing circumstances (Slabbert et al. 2010). According to Knoppers et al. (2007), there has been a move in biomedicine from genetic to genomic research, which operates at the level of the whole genome, and studies 'normal' genomic variations across whole populations (Knoppers et al. 2007, p. 291).

Functional genomics aims to characterize the many different genes that constitute these genomes and their variability of action (WHO 2002, p. 4). Pharmacogenomics aims to develop therapeutic interventions individually tailored to the biochemical makeup of the patient. Monogenic diseases resulting from mutations from a single gene, such as the most common forms, Thalassemia and sickle cell disease, provide particular challenges as they are generally incurable and require a lifetime of medical treatment.[7] Techniques have been developed for

[7]There are about 5000 diseases classified as monogenic, such as inherited haemoglobin disorders, cystic fibrosis, haemophilia, with the global prevalence being about 10 per thousand at birth (WHO 2002, p. 43).

carrier detection, which are incorporated into the public health system. Most common diseases, heart disease, stroke, diabetes, the major psychoses, result from more complex environmental factors as well as the effect of ageing, together with variations in individual susceptibility reflecting the actions of several genes. Public health genomics has thus emerged as a distinctive field, which grapples with the challenge of positioning the practice and applications of genomics so as to best advance medical science and public health (Brand et al. 2008, p. 5). Pathogen genomics is another related branch of medical science that studies pathogens—viruses, bacteria, parasites—and their vectors, in order to understand the dynamics and ecology of infectious diseases.

Both predictive and preventative knowledge is sought in all these interlinked fields. Genome-Wide Association Studies (GWAS) map complex disease susceptibility and leverage the composite information in order to test for disease association. Genomic variation worldwide is explored in studies such as the International HapMap Consortium,[8] a project that is in the process of developing a haplotype map of the human genome that describes the common patterns of DNA sequence variation (International HapMap Consortium 2005). *Haplotypes* are genetic markers, which are used to seek patterns of genetic variation in order to illuminate the linkages between genetics and disease.[9] In 2008, an international consortium announced the 1000 Genomes project, a major human sequencing effort aiming to produce the most detailed map of human genetic variation in order to support disease studies.[10]

In the words of Richard Durbin of the Wellcome Trust Sanger Institute, "such a project would have been unthinkable only 2 years ago. Today, thanks to amazing strides in sequencing technology, bioinformatics and population genomics, it is now within our grasp" (NIH 2008, p. 4). The choice of populations for genome sequencing targets, those with particular ancestry, are described as,

> Yoruba in Ibadan, Nigeria: Japanese in Tokyo, Chinese in Beijing: Utah residents with ancestry from Northern Europe, Luhya in Webuye, Kenya: Maasai in Kinyawa, Kenya, Toscani in Italy, Gujarati Indians in Houston, Chinese from Denver, people of Mexican ancestry in Los Angeles, and people of African ancestry in the United States. (NIH 2008, p. 3)

[8]The HapMap project website is http://www.hapmap.org and publishes up to date information on the status of the project.

[9]Haplotypes are defined on the Human Genome Project website as, "groups of closely linked alleles that tend to be inherited together … and which can be used to map human disease genes very accurately". [Online] Available at http://www.wellcome.ac.uk/en/genome/thegenome/hg04b002.html.

[10]The project is comprised of the Wellcome Trust Sanger Institute in England, the Beijing Genomics Institute in China, and the National Human Genome Research Institute (NHGRI) part of the National Institutes of Health, America. [Online] Available at http://www.1000genomes.org (NIH 2008).

1.2 Human Population Genetic Research

As Sheremeta and Knoppers (2007) point out, human population genetic research (HPGR), in which the DNA from thousands of research subjects, is linked to medical records and genealogical data and seeks to identify susceptibility genes for common diseases, is viewed as a 'next step' in the evolution of research based on the human genome. Databases containing medical and genealogical information resulting from such research are anticipated to provide insight into etiology and prevention of complex human diseases (Sheremeta and Knoppers 2007, p. 157). It is beyond the scope of this book to describe the entire evolving field further. Fujimura comments on how a reductionist genetics of the past 40 years has given rise to new fields in a post genomic era, namely systems biology and development systems theory, which attempt to grapple with the complexity of life (Fujimura 2005, p. 196).[11]

Uses of the above have powerfully entered public health systems of developed countries. For example, susceptibility genes have been discovered for a number of diseases such as Crohn's disease and type 2 diabetes, which enable identification through genetic testing (WHO 2002, p. 56). Certain populations are more susceptible to particular diseases, for example the Ashkenazi (Eastern European origin) Jewish population was discovered some 30 years ago to be significantly more prone than others to be stricken by two inherited and fatal conditions, namely the Tay-Sachs and Canavan diseases.

Approximately one in thirty individuals carried the former gene, and one in thirty-eight the latter. Research initiated by family groups who worked with geneticists led to the establishment of effective genetic screening of individuals, which has resulted in isolating the genetic causes not only of the above two diseases, but of nine others, including Familial Disautomia, Gaucher disease, and Cystic Fibrosis (Levene 2004). A New York Times article titled *Using Genetic Tests, Ashkenazi Jews Vanquish a Disease* praised the collaborative efforts that have resulted in the virtual elimination of the Tay Sachs and Canavan diseases amongst the vulnerable population, quoting a certain Doctor Schriver as saying, "the Tay-Sachs and Thalassemia carrier screening programs over their 30-year existence in Montreal have resulted in an almost complete absence of new cases of these two diseases" (Kolata 2003, p. 1).

Some branches of HPGR are non-medical and non-commercial in their focus, which does not make them immune from controversy. Population genomics is a broad field that carries out studies of normal genomic variation across whole populations, collecting bio-samples and data on a longitudinal scale (Khoury 2004). Comparing sequences of groups from different continents allows scientists to define relationships and the ages of different populations. Public ancestry projects benefit

[11]These new fields of study include Rnomics (for RNA) proteomics, Systeomics and Physiomics. (Fujimura 2005, p. 197).

from this research, enabling individuals to trace the continental origins of their ancient ancestors. By combining genetic data with archaeological and linguistic information, anthropologists have been able to discern the origins of *Homo sapiens* in Africa, and to track the timing and location of the migrations from Africa that led to the eventual spread of humans across the world (Jones 2009, p. 30).

The Human Genome Diversity Project (HGDP) conceived in 1992 was an ambitious global research project to investigate human genetic diversity, and to survey variations in the human genome across the entire human population (Roberts 1991). Genome-wide association studies were to be carried out to study human genetic variation and common complex diseases, using the alternative category of *genetic ancestry* rather than former notions of race and ethnicity for genetic analysis, which led to the use of the term *genome geography* within this branch of research (Fujimura et al. 2011, p. 5).

The misunderstandings and political predicaments that arose were an indication of the potential sensitivity attached to the entire field of genomic research, and a caution to those that enter the field. As MacIntosh put it,

> Genetic research involving indigenous populations provokes many legal, ethical and cultural issues. Like any genetic research project, it incites questions about ownership of genetic samples including the information gleaned from those samples and, of course, about whether human genetic material is or ought to be patentable. (MacIntosh 2005, p. 214)

Fears by many of the discrete groups selected by the project for inclusion by reason of their being unique, historically vital populations in danger of dying out, led to attacks on the HGDP becoming so virulent in nature that it acquired the sensational title amongst its many critics of the "vampire project" (Lock 1997, p. 79). The Indigenous Peoples Council on Biocolonialism (IPCB) was formed in 1993 in response to the HGDP, in order to create an active voice on behalf of indigenous peoples.[12] At least ten collective statements were subsequently made by indigenous peoples rejecting the HGDP, including the Mataatua Declaration on Cultural and Intellectual Property Rights of Indigenous People of June 1993, which was formulated by 150 indigenous participants from 14 member states (Thambisetti 2002a, p. 21).[13]

Seeking to overcome the negative press associated with the HGDP, the Genographic Project[14] was launched by National Geographic and IBM in 2005 with its primary aim to gather DNA samples from 100,000 indigenous peoples across the globe in order to try and track human migratory pathways over past

[12] [Online] Available at http://www.ipcb.org.

[13] Other declarations including the World Council on Indigenous Peoples, Maori Congress Indigenous Peoples Roundtable of June 1994: Latin and South American consultation of Indigenous Peoples Knowledge, September 1994: Declaration of Indigenous Organisations of the Western Hemisphere, February 1995: Asian Consultation on the Protection and Conservation of Indigenous (Peoples Knowledge, February 1995: National Congress of American Indians, Resolution No NV 93-118): These and more are listed on the website of the Indigenous Peoples' Council on Biocolonialism, http://www.ipcb.org.

[14] The Genographic Project official website is http://www.nationalgeographic.com/genographic.

millennia. Clearly, much trust had been lost in previous processes, for as stated by Fujumura et al. it is the population differences in the field of genetic research, especially where populations are selected because of their genotypic properties that have been shown to have the potential to evoke discord (Fujimura et al. 2011).

Whilst some authors have argued the extreme position that the likelihood of indigenous peoples being exploited is so high that they should generally resist participation or involvement in genetic research,[15] an emerging and more balanced view is that with appropriate consideration of benefit sharing in the interaction, just relationships can be formed between researchers and communities in furtherance of genetic research (MacIntosh 2005, p. 215; Sheremeta et al. 2004, p. 89). According to Wang, the HGDP which targets specific isolated, indigenous, and often vulnerable, peoples is an important form of human population genetic research (HPGR) exposing those groups to particular concerns and distinctive risks which justify better forms of regulation and protection (Wang 2011, p. 14).

1.3 Population Bio-Banks

The term *population bio-banks* is used to describe large repositories of human DNA or gene banks from which genomic databases can be derived, and which may include extracted DNA, body fluids, cells, sections of tissue, and information encompassing molecular genetic data.[16] Bio-banks form novel globalized research infrastructures that hold special promise for research and clinical medicine. Whilst genetic registers have existed for much of the past century, since the Second World War they became focused primarily on research and on providing genetic services to people with particular genetic diseases (Hook 1976). The first significant attempts to create ambitious national bio-banks for population-level genomic studies, such as those in the United States, Iceland, the United Kingdom, Latvia, Estonia and Sweden share ambitious aims to further genomic research. These have thrown up new ethical and legal challenges relating to issues of consent, property rights, access to data, privacy and protection against discrimination that are subject to much academic scrutiny (Williams 2005; Winickoff 2003; Chadwick 2011; Simm 2005; Arnason 2011).

---

[15]Debra Harry *IPCB Action Alert to Oppose the Genographic Project* http://www.ipcb.org/issues/human_genetics/htmls/action_geno.html>Harry.

[16]Population bio-bank is defined in A 17 as, "a collection of biological materials that has the following characteristics: (i) the collection has a population basis (ii) it is established, or has been converted, to supply biological materials or data derived there from for multiple future research projects (iii) it contains biological material and associated personal data, which may include or be linked to genealogical, medical and lifestyle data and which may be regularly updated, and (iv) it receives and supplies materials in an organized manner". (Biobank Lexicon: Public Population Project in Genomics and Society. http://www.p3g.org/biobank-lexicon, Article 17).

Whilst many developed world bio-banks make it clear to research participants that they have no property rights in their bio-samples or data (Uranga 2005, p. 62), and that they should have no expectation of personal financial gain,[17] Knoppers et al. (2007) provide a reminder that there is currently no single mandated form of international institutional control over population bio-banks (p. 294). They also make the important point that no bio-banks have excluded the possibility that some research may lead to biomedical products that return a profit, which can function as sources of income, and of re-investment in the databases (ibid, p. 302).

The question of the most appropriate distribution of risks as well as benefits across the entire spectrum of stakeholders that will be described subsequently as the *legal* neighbourhood, which includes researchers, human research participants, funders, medical institutions and private sector partners, has given rise to a wealth of academic commentary. A recent study of bio-banks singled out consent, confidentiality and commercialisation as the central and enduring issues requiring development and consensus (Knoppers et al. 2008).

Concerns noted by Williams and Schroeder include the fact that whilst founding partners of such bio-banks determine the research priorities, their commercial partners are well placed to exploit sample collections in these bio-banks selectively; even if some benefits do emerge, such exploitation should be regarded as inequitable (Williams and Schroeder 2004, p. 100). The associated issue of eventual commercialisation flows from the question of ownership of tissue samples, or of data derived there from, which has been treated in different ways by different bio-banks. For example, the Icelandic Acts on Bio-banks explicitly provided that the bio-bank was not to be considered the owner of the biological sample (Boggio 2005, p. 46), whereas the UK Bio-bank is the legal owner of the database and the sample collection, where participants explicitly do not have property rights in the samples.[18]

An additional complication is the fact that ownership of a DNA sample is separate and distinct from the issue of ownership of, and intellectual property rights to, commercially viable products derived from the information contained in that sample, which is seldom explicit in bio-bank policies (Andanda 2008, p. 173). Whilst the right of control over one's body based upon notions of personal autonomy is well established, a property approach to human tissue is stated by many as being legally unworkable (Knoppers et al. 2008, p. 11) and is discussed together with aspects of commercialisation below.

---

[17]The UK Bio-bank Ethics and Governance Framework provides in Section 1(B) 1 under 'Consent' that, "UK Bio-bank will be the legal owner of the database and sample collection, and participants will have no property rights in the samples", and under Section 1(B) 8 that participants may have no expectation of personal financial gain. [Online] Available at www.ukbiobank.ac.uk/wp-content/2011/05/EGF20082.

[18]UK Bio-bank: Frequently asked Questions [online] Available at http://www.ukbioibank.ac.uk/about/faqs.php#anonymous.

1.4 Global Biomedical Research

Knowledge is stated by some to be the archetypal public good (HUGO 2000) as evidenced by phrases such as, "research produces knowledge that serves the Greater Good" (Parry 1986, p. 486). Global research seeks to contribute towards this broad aim of knowledge for all, this higher purpose frequently made explicit by use of terms such as the common good or the public interest (Simm 2011, p. 556). The aspirational notion of a right to health, first incorporated into international law in article 25(1) of the Universal Declaration of Human Rights (UN 1948), was followed and strengthened by article 12 of the binding International Covenant on Economic, Social and Cultural Rights (ICESCR 1966) of 1966, which enjoins states to strive for this objective which underpins much of the global research endeavour.[19]

Genetic and genomic research, often associated with public and private bio-banks, constitutes an increasingly important component of global research, not only changing the landscape of possibilities, but also influencing the field and discourse in research ethics (O'Neil 2004). It is often argued that a major tension in bioethics is between protecting the private interests of individuals on the one hand, and contributing towards the common good on the other (Arnason 2011, p. 563). The *common good* is a phrase with numerous relatives such as *public interest*, *common interest* and *public good* all of which attempt to capture an essence of public rather than private good (Simm 2011, p. 554).

Contemporary population research bio-banks are designed to accomplish indefinite linking capacities with associated research infrastructures, resulting in what Karlsen et al. describe as a quantitative leap in terms of data. For example, the 1000 Genomes Project anticipated generating 60-fold more sequence data in 2 days than had been added to public data bases over the past year (NIH 2008, p. 3). Whether this enhanced research capacity will however facilitate an epistemic leap of equal size, expanding humankind's scientific and medical knowledge, remains to be seen (Karlsen et al. 2011, p. 574).

The global advance of medical science depends upon the free flow of data between numerous players, including scientists, academic institutions, multinational corporations and states, and the manner and degree of access to research data remain of central importance. Non-profit research organisations, small specialized companies, and government agencies all interact and play roles, with research benefits as well as public and private finances freely crossing international borders. For example, the long-enduring research project on the Majengo sex workers of Kenya has involved at least three universities (Manitoba, Oxford, Nairobi) two of which are from the developed world (Canada, UK) as well as the Kenyan Medical

[19]Article 12 of ICESCR states, "Parties to the present Covenant recognize the right of everyone to the enjoyment of the highest attainable standard of physical and mental health". (ICESCR 1966, p. 7).

Research Institute. The 1000 Genomes Project referred to earlier which links research institutes from China, the UK and the USA is another example of the globalised nature of research. These international collaborations provide a virtual location for the research endeavour, defined and constituted by transferability, capacity, efficiency and accessibility of information (Karlsen et al. 2011, p. 574).

1.4.1 Funding and Collaborations in Research

According to the World Health Organisation, there is a danger that global spending on health research, already skewed at a 90:10 ratio in favour of the developed world, will become more skewed by the genomics revolution, and also that private company spending on genomics is already substantially higher than that from public sources. The concern is thus that the central role of entrepreneurial or risk capital will result in research priorities being driven purely by market considerations and the profit motive, aimed at satisfying demand for diseases prevalent in the developed world (WHO 2002, p. 127). For example, the United Kingdom alone spent over $200 million on cancer research in 1993,[20] whilst global research on Malaria, which kills over a million children a year, received a mere $84 million (Macklin 2004, p. 9). Population genetic research is by its nature truly global in its reach, and population databases require a high level of capital investment.

Whilst public bio-banks are largely funded from public sources, for example, the UK Bio-bank by the Wellcome Trust, the Medical Research Council and the Department of Health (Bio-bank UK, Online) and the H3Africa project by the NIH (NIH 2012), others are funded by partnerships often involving private companies and risk capital. For example, DeCODE Genetics Inc entered into a partnership with the Icelandic government for the Icelandic Health Sector database with an exclusive license to exploit the database for 12 years (Greely 2000). The Estonian Gene Bank Project (EGP) a not for profit organisation founded by scientists, physicians and politicians, provided an exclusive license to the corporate EGeen, a US for-profit entity, with the aim of exploiting intellectual property rights in drugs and diagnostic targets emerging from the project (Habeck 2002).

Similarly, the Tonga population database project was meant to be a collaboration between the Tongan Ministry of Health and an Australian biotechnology company, Autogen (WHO 2002, p. 116). The private sector is thus deeply implicated in the global research venture, fuelled by the drive to commercialise the results of research that is often funded from public sources (Caulfield and Chapman 2005). Many universities form spin-off biotechnology firms in order to benefit from their research (Statistics Canada 2002), as scientific developments in the fields of biotechnology,

[20] All dollar amounts in this book refer to US Dollars unless otherwise specified.

biochemistry, molecular biology, cell biology, immunology and information technology as well as genetics and genomics contribute towards a lucrative field of drug discovery and development.[21]

The sequencing of the human genome opened a new world of genetic biomarkers to be used in the search by pharmaceutical industry for new drugs (The Economist 2008), which development was described by some as the "corporatisation of research" (Rajan 2006, p. 4). Genetic data are in many cases in and of themselves potentially lucrative, even before any medical products have been developed. For example, *Incyte Parmaceuticals* reported $220 million in profits in 2001 for merely selling access to genes on which it held, or had applied for patents (MacIntosch 2005, p. 217).

Collaboration and integration between companies is also changing. During 1997, no fewer than 374 agreements were reported to have been concluded between large pharmaceutical firms and small drug discovery companies. Moreover, more than half the substances in clinical trials originated outside the laboratories of the biggest pharmaceutical companies (Laird et al. 2002, p. 248). However, where multinational pharmaceutical companies traditionally dominated the field, nowadays mergers and strategic partnerships with smaller innovative biotechnology, biopharmaceutical or drug discovery companies are increasingly common in the rapidly changing industry (Economist 2008, p. 57). A survey of the pharmaceutical industry in 1998 had held that US and UK based firms allocated roughly 20 % of their estimated $300 million sales per annum to research and development whilst costs of developing a new drug, the period for which is between 8 and 10 years, were estimated at over $500 million (Economist 1998).

A more recent analysis by Herper (2012) compared the twelve largest pharmaceutical companies' spending on research over the past 15 years. He found that *AstraZeneca* at the top end had spent $11 billion for every new drug approved, whilst *GlaxoSmithKline* had spent $8,1 billion, *Pfizer* had spent $7.7 billion, and *Amgen* had spent just $3.7 billion per new drug. In an in-house report of the Pharmaceutical industry entitled The Big Pharma Recession Report at least 71 % of companies predicted that the global recession was bringing about massive structural changes in the industry, and that the cost of bringing a drug to market was reported as having risen to $1.048 billion in 2011 (Pharma 2012). This cost per new drug shows a remarkable disparity with the $4–$12 billion dollars calculated by Herper above.

A sobering statistic to place this global drug research enterprise into context is the fact that out of thousands of compounds screened, few are likely to become an approved drug. One estimate states that for every 10,000 molecules screened, an average of 250 enter pre-clinical testing, ten make it through to clinical trials, and only one is approved by the regulator (The Economist 2008). For another example, the Drug Discovery and Development Centre of the University of Cape Town

[21]In 1999, three of ten biotechnology firms in Canada were spin-off companies, have with 91 % were formed by universities or teaching hospitals. (Statistics Canada 2002).

spent 18 months analysing over six million chemical compounds, narrowed the search to 15,000 active compounds for use against malaria, and finally announced that it had found one, named MMV 390048, which would enter clinical trials (Price 2012, p. 5).[22]

New ways of sharing data, conducting research and forming cross-disciplinary teams thus affect the path of contracts, titles, reward and intellectual property, in order to deal with, as a *Time* lead article described it, "the paradigm shift necessitated by the torrent of data pouring forth from genomics" (Saprai 2009, p. 22). For example, NIH head Dr Francis Collins currently leads no fewer than 27 institutes collaborating in projects exploring genomics and gene expression in the study of cancer, grouping all manner of disciplines in the joint quest (ibid, p. 23). The transformation of institutional research is required in order to,

> [...] take advantage of the dazzling scientific and technological advances that have taken place in just the past 3 years—advances in bioengineering, nanotechnology, drug compounds and data gathering, including protein data, splicing data, and mutation data, all hoisted into view by ever cheaper computational muscle (ibid, p. 24).

Progress in genome sequencing has led to pharmaceutical leaps, hundreds of drugs are in development for therapies targeting identified genetic mutations (ibid, p. 24), and the search for drug discoveries remains at the epicentre of global health research.[23] Within this wide spectrum of possibilities, multiple stakeholders collaborate in ever more novel ways, involving a large number of possible alliances between state, public donor, university, researcher, private institution and various levels of community. Amongst these multiple stakeholders, indigenous and other vulnerable peoples are of particular interest, given that they are often the target of HPGR, and in view of the fact that exploitation claims (see below) are frequently raised by their representatives or those speaking on their behalf in the context of genetic research.

1.4.2 Indigenous Peoples in Genetic Research

The term *indigenous peoples* referred to above alongside *Aboriginal peoples* and related descriptions is frequently used in discussions on HPGR and other genomic research in the developing world, and calls for a brief explanation. Whilst the term has become part of the United Nations lexicon, no United Nations body has ever adopted a definition. The International Labour Organisation had defined the term

[22]The UCT Drug Discovery and Development Centre, H3D, is a consortium of Griffith University of Australia and Medicines for Malaria Ventures (MMV), which is funded by the Bill and Melinda Gates Foundation.

[23]Sequencing the first human genome took more than a decade and an estimated $2.7 billion, whilst today sequencing of a human genome can be done for a few thousand dollars in a few hours.

in its Convention 169 Concerning Indigenous and Tribal Peoples in Independent Countries, as those peoples,

> […] who are regarded as indigenous on account of their descent from the populations which inhabited the country, or a geographical region to which the country belongs, at the time of conquest or colonisation or the establishment of present state boundaries and who, irrespective of their legal status, retain some or all of their own social, economic, cultural and political institutions. (ILO 1989, Article 1)

The United Nations General Assembly has subsequently placed indigenous peoples firmly on the map as a significant and identifiable grouping in world affairs, and after declaring two official decades of indigenous peoples commencing in 1995[24] as well as creating the Permanent Forum for Indigenous Issues (UNPFII) in 2000.[25] It finally adopted the Declaration on the Rights of Indigenous Peoples in 2007, describing the document as setting, "an important standard for the treatment of indigenous peoples that will undoubtedly be a significant tool towards eliminating human rights violations against the planet's 370 million indigenous people" (UNDRIP 2007).

Indigenous peoples are frequently referred to as epitomising vulnerable populations in need of particular levels of protection from exploitation (Chennells 2009; Weijer 1999; Weijer et al. 1999), a recurring theme being the concern that they tend to have a weak bargaining position due to poverty and associated factors, making them particularly vulnerable to exploitation (Dutfield 2004, p. 48). The fact that they are generally lacking in literacy as well as experience regarding engagement with the modern world has given rise to much literature on the complexity of obtaining informed consent from such communities not only in the CBD (the Convention on Biological Diversity) context, but also in genomic as well as other forms of research (Upshur et al. 2007; Weijer 1999, 2004; Kegley 2004).

Utilisation of the traditional knowledge or genetic resources of indigenous peoples without having obtained their prior informed consent was a central component of the allegations and perceptions of exploitation that influenced the formation of the CBD, and the issue remains equally relevant for genomic research. Both of these topics receive further attention below. Indigenous peoples, frequently sought after for participation in genomic and other scientific research, have in recent years begun to articulate their own concerns more assertively, recent examples of which are the IPCB described above, and the Waitangi Tribunal of New Zealand, a legal commission which published the beliefs and policies guiding bioprospecting and research on the Maori peoples (Waitangi 2011). In Australia genetic research with indigenous communities has been the focus of much attention (Kowal 2012), and in Africa the interests of indigenous peoples are represented by the continent-wide organisation Indigenous Peoples of Africa Coordinating Committee (IPACC).

The cosmology shared to a large degree by indigenous peoples has frequently resulted in what Dutfield terms the politicising of issues such as commercialisation of research (Dutfield 2004), as reflected in the following extract from a statement

[24]Proclaimed by General Assembly resolution 48/163.

[25]Proclaimed by UN Economic and Social Council resolution 2000/22 [2] on 28 July 2000.

issued in opposition to the prevailing intellectual property system: "patenting and commodification of life is against our fundamental values and beliefs regarding the sacredness of life and life processes, and the reciprocal relationship, which we maintain with all creation" (Tauli-Corpuz 2003, p. 25). Indigenous peoples provide an important source of material for genomic research, as will be further explained during this book. They have, however, experienced the demand for their DNA with considerable mistrust, caused not only by cultural and other misunderstandings but also influenced by their history of domination by more powerful peoples. Their particular concerns which include mistrust of research objectives, the fairness of the exchange, as well as the collective risks associated with genetic research justify better forms of regulation and protection than that presently available (Wang 2011, p. 14). These concerns will be addressed during the book and particularly in Chap. 8 below.

1.4.3 Setting Research Priorities

As is now apparent, a significant proportion of global research is funded by the private sector, which in most cases formulates the research priorities and seeks a commensurate return on investments. States also remain critical role players in global research, not only as the locus of domestic legislation and policy regarding scientific research and the public health, but in addition as funders of universities, bio-banks and specific research projects. They are encouraged in this regard by UNESCOs' Universal Declaration on Bioethics and Human Rights, which holds that states should,

> [...] promote equitable access to medical, scientific and technological developments as well as the greatest possible flow and the rapid sharing of knowledge concerning these developments and the sharing of benefits with particular attention to the needs of developing countries. (UNESCO 2005, Article 2f)

Whereas human population genetic studies are largely focused on finding cures for diseases, the arrangements between research bodies and other stakeholders describe and govern the particular research objectives, including possible patenting and commercialisation of innovations, and also determine the advantages or benefits that accrue to all parties. Concerns have been raised about many aspects of these research arrangements, including the fact that research objectives are predominantly driven by the commercial viability of the intended health discovery. All of these concerns emerged in the objections against the Human Genome Development Project (HGDP) discussed further below, which was an attempt to research selected indigenous peoples worldwide.

The HGDP erupted in controversy during which indigenous peoples demanded a role in defining research agendas, as well as in interpreting the facts and accruing the monetary and other benefits of the research itself (Reddy 2007, p. 442). In the words of Brodwin (2005), this resulted in a new politics of recognition that affected global research and rendered human population genetics forever politically vulnerable (ibid, p. 148). The concerns expressed during these and subsequent debates included the fact that research priorities are driven by the health aspirations of the relatively affluent developed world, ignoring the more urgent health needs of developing countries (Williams and Schroeder 2004).

An additional and serious level of unfairness is entailed in the allegation that in the process of genetic research across borders, large profits are made from research on human tissue whilst the contributors remain unaware of and barred from sharing in the benefits (Annas 1990). With so much research going on, with so many stakeholders and organisations joined in the race to discover valuable properties of our genetic makeup, it is not surprising that a range of problems have reached the public domain over the years. Most concerning are the continuing perceptions and allegations of unfairness or exploitation which have followed such genetic research.

1.5 Perceived Unfairness in the Exchange of Human Genetic Samples?

Within the field of international genomic research, a number of cases have raised questions and given rise to uncertainty or public concerns.[26] The avian flu crisis that erupted in Indonesia in 2006, where the first human infections had occurred, centred on allegations of unfairness and provided controversy on a global scale. The Indonesian government had decided to withhold further virus samples from the WHO after it discovered that an Australian company had applied for a patent based on earlier samples, complaining about the unfairness of the system involving the exchange of virus samples which enabled private companies and developed countries to exploit and benefit from the development of vaccines, whilst Indonesia remained excluded. It claimed sovereignty over its genetic resources as the basis of its stance, and rejected the then prevailing system as being exploitative and unfair (Sedyaningsih et al. 2008).

Amongst the arguments raised by the Indonesian government was firstly, the assertion that the CBD provides states with sovereignty, thus ownership over all its biological and genetic resources (ibid, p. 485). Secondly, that this gave Indonesia a right to 'prior informed consent' over its resources, and thirdly, a 'reciprocity' argument pointing out the extreme unfairness in the exchange (ibid, p. 487). The controversy resulted in negotiations towards an entirely new virus exchange system under the auspices of the World Health Organisation which was concluded in 2011, and addressed the concerns of developing world states that had previously felt exploited while balancing the interests of researchers and innovators (Wilke 2012, p. 12).[27]

Controversy has dogged many cases of human genomic research, in most cases involving selected groups of relatively isolated research participants described

[26]Some cases have become infamous as examples of wrongful exploitation, such as the Tuskegee studies on untreated syphilis in poor African American men in rural Alabama (Angell 1997). In addition to this is the case of Henrietta Lacks where cancer cells from a terminally ill black woman were obtained in order to develop highly successful cell lines which were then used freely in research for over 25 years without the knowledge or consent of her or her family (Skloot 2010).

[27]Resolution adopted by the World Health Assembly on 24 May 2011. Document 64.5 entitled Pandemic Influenza Preparedness: Sharing of influenza viruses and access to vaccines and other benefits.

variously as indigenous, Aboriginal, rural or tribal peoples from the developing world. For example, the curious genetic history of the islanders of Tristan da Cunha had made them the objects of much genomic and other research over the years, for the reason that the approximately 275 inhabitants were all descended from no more than fifteen original families at the time of settlement in 1816. Their genealogy has been well documented (Soodyal et al. 2003) but it was an academic research project by a University of Toronto affiliate (MacIntosh 2005, p. 218) entitled, *In search of the genes of asthma on the island of Tristan da Cunha* following discovery of the high incidence of asthma on the island that led to a race to discover the genetic cause of asthma (Zamel 1995, p. 6; Zamel et al. 1996).[28] The research institute worked in conjunction with a genomics company *Sequana Therapeutics*, which promised to provide islanders contributing their DNA samples free pharmaceutical treatment in the future if drugs were ever to be developed based on their research (MacIntosh 2005, p. 218).

Sequana Therapeutics announced in 1995, that it had identified the gene, which predisposed people to develop asthma, and then promptly sold the rights to develop a genetic diagnostic test to a German company, *Boehringer Ingelheim*, for $70 million (Cunningham 1998, p. 18).[29] Many commentators criticised the vast disparity between the realized benefits in the hands of *Sequana* and *Boehringer* and the potential future benefits to the islanders who would only receive free pharmaceuticals if and when a drug based treatment were ever to be developed (Cunningham 1998; MacIntosh 2005; Hamilton 2001). The case bore all the hallmarks of the claims of injustice relating to the exchange of non-human genetic resources that had led to the negotiation of the benefit sharing provisions of the CBD, namely the free provision of genetic material from naive developing countries to commercial entities, who then stood to make their profits downstream with no compensation offered to the original source.

Many genomic research cases that resulted in legal challenges revolved around the nature, content and validity of the informed consent that preceded the provision of human tissue. The Havasupai Indians living in the Grand Canyon sued the University of Arizona for recovery of samples taken from the tribe in the early 1990s, claiming that the samples were being used many years later for unknown purposes beyond and outside the terms of the original consent. The University claimed no wrongdoing in pursuing research leads which may have led to important and lucrative discoveries without reference to the source of the samples (Beauchamp 2011, p. 520). After protracted legal action a settlement was reached in which the University agreed to return the blood samples, provide other forms of assistance to the tribe, and to pay damages in an amount of $700,000 to 41 of the tribe's members (Harmon 2010, p. 1).

One of the central factors emerging from the court papers was the fact that blood has a deep spiritual meaning to the Havasupai, which had been ignored in the research procedure (ibid, p. 3). The case was paradigmatic in its exposure of the difficulty of obtaining ethical, and legally valid, consent. Not only was the

---

[28] The Samuel Lunenfeld Research Institute of Canada.

[29] A BBC press release entitled '*Worldwide search for a cancer cure*' described the asthma gene as "ESE3" [Online] Available at http://news.bbc.co.uk/2/hi/health/7766656.

broad consent obtained by the university inadequate and inappropriate, but also the university committee review of the research had proved ineffective (Beauchamp 2011, p. 520). Amongst the many disturbing issues raised by the case is the fact that the university researchers and review committee had all overlooked or failed to perceive the serious risks involved for the tribe, including discrimination and disrespect for cultural traditions, raising questions of exploitation and taking advantage of a vulnerable group (ibid, p. 521). This issue of the particular cultural and identity-related risks facing such indigenous groups when invited to participate in genetic research will be developed further in Chap. 8 below.

In a similar matter, the Nuu-cha-nulth tribe of Vancouver Island in Canada publicly criticized the University of British Columbia for sharing human genetic samples that had been taken from them over 25 years previously, specifically in order to investigate rheumatoid arthritis with a variety of other research institutions for varying purposes (Dalton 2002). The research exposed the spread of lymphotropic viruses by intravenous drug use, stigmatising the tribe in the public arena. The Yanomami Indians of Brazil likewise claimed that researchers had misused their samples, a well-publicized saga that was documented in a book by Tierney entitled, *Darkness in El Dorado: How scientists and journalists devastated the Amazon* (Tierney 2000). In both of the latter two cases, the controversy had revolved not only around the nature and content of the original consent, but in particular around the subsequent unauthorised use in the developed world of their human tissue samples.

Other claims of exploitation by indigenous peoples have carried a more directly commercial flavour. For example, following genetic research on indigenous Guaymi tribespeople from Panama, the Secretary of Commerce of the US government laid patent claim to a cell line created out of the blood taken from a 26 year old indigenous Guaymi woman who had leukaemia. If granted, this patent would have given the US government exclusive rights to decide who could use this potentially valuable cell line, and at what cost. Widespread objection was lodged against the patenting and commodification of the provided tissue without the tribe's or the woman's knowledge or consent, leading to renunciation of the patent (Christie 1996).

In yet another case, the Hagahai tribe of Papua New Guinea sought the help of a geneticist to discover why so many of their children were dying of a certain illness. Researchers found that the members of the tribe carried a certain gene, which predisposes humans to Leukaemia, whilst they remained immune to the sickness. Further analysis of the blood identified a rare virus with potential for development into a vaccine for certain types of Leukaemia. An agreement was then reached between the United States National Institutes of Health (NIH) and the tribe that they would share 50 % in the royalties of any breakthrough in Leukaemia research resulting from their genetic sample (Salopek 1997, Part 1).

A patent was registered for a cell line drawn from a Hagahai donor, which was the first to be based on a human cell line taken from an indigenous population (Lehrman 1996, p. 500), leading to widespread claims of biopiracy flowing from the alleged lack of informed consent (Friedlander 1996). Controversy erupted about the case, despite the fact that the tribe appeared to have consented to the research, with human rights groups nevertheless alleging injustice and unfairness. The core of the criticism focussed on the taking of a valuable item from an illiterate tribe from a poor country without acknowledgement of the commercial potential. "What government has the right to poach on a foreign country's genetic heritage?" was the statement by human rights group Rural Advancement Fund International (Salopek 1997, Part 2).

The patent, which according to the NIH identified unique genetic traits relating to Leukaemia and was in every way fully justified, was nevertheless eventually revoked by this body in the face of the widespread international outrage (Tokar 2002). Indigenous peoples as well as supportive civil society organisations responded in a manner reflective of the anger addressed at the HGDP referred to above. One Aboriginal leader John Liddle expressed himself as follows, "Over the last 200 years, non-Aboriginal people have taken our land, language, culture, health, even our children. Now they want to take the genetic material that makes us Aboriginal people as well" (ibid, p. 3).

A succession of other cases involving remote populations that received prominence in DNA research are listed in a series of reports by award-winning journalist Paul Salopek.[30] The village of Limone Sur Garda in Italy, having been cut off from the road system and thus from civilisation until the 1950s, developed a particularly untouched gene pool regarded as valuable for that reason. Geneticists discovered a mutation in the villagers' DNA, which appeared to render them immune to atherosclerosis, leading to their being heavily researched in pursuit of a cure. The Pima Indians of Arizona, a 6,000-strong tribe of former hunter-gatherers, developed health problems with the switch to a more sedentary lifestyle, including the highest diabetes rate of any ethnic group in the world. NIH researchers traced genealogies and researched the Pima for 30 years, trying to discover the genetic causes of a disease that affects millions worldwide.

Oklahoma's Cherokees were researched for their remarkable resistance to Alzheimer's. Israel's Bedouin were being studied for clues to obesity and certain inherited forms of deafness, and the Amish for their apparent high rates of inherited depression (Rowe 1996, p. 1320). A French biomedical firm *Genethon*, discovered a remote population on Reunion Island in the Indian Ocean that appeared to be immune to multiple sclerosis (MS), even though they carried the disease genes. The book driving this kind of research was that the protective genetic factor might, if isolated, have led to lucrative new drugs for MS sufferers, making this population potentially of great value for the developed world (Salopek 1997, Part 2). Many other such cases are recorded (Tokar 2002).

As a prominent example of the urgency of the biomedical industry's interest to discover genes leading to lucrative drugs, the international biomedical company *Genset* signed a deal in 1998 to sift China's 1.2 billion citizens for mutations, focusing on remote rural populations and its approximately 40 minority groups. The contract to prospect for lucrative genes in these populations was regarded as breath-taking in its extent, and of incalculable but immense value (Salopek 1997, Part 3). Others have explored Chinese genes for a clue to the remarkable longevity of certain communities (Dang et al. 1999). A vigorous debate on the ethics of

---

[30]Salopek won the Pulitzer award for explanatory reporting in 1998 for his article, *Genes offer sampling of hope and fear*, in the Chicago Tribune of 28 April 1997.

research in China included claims of exploitation of vulnerable populations by the Harvard researchers, one issue being whether Harvard research and consent protocols were appropriate for communities with desperate medical needs, or who culturally fear offending those in authority (Rose 2004, pp. 7–8). Finally, an article by Brian Tokar published by the Third World Network entitled *Mining humanity* describes how scientists have targeted tribes people in Saudi Arabia in search of genes linked to glaucoma; in Ghana and Nigeria to study diabetes; in Mongolia for studies on congenital deafness, and in remote Philippine islands to study the unusually high incidence of cleft lip and palate (Tokar 2002, p. 1).

These cases have been described in some detail in order to highlight the concerns being addressed by this book, and which have alerted remote and indigenous peoples of the developing world to pay careful attention to the purposes of intended research, as well as to the details of the consent given (Pellekaan 2002). It is without doubt that the DNA of such remote and indigenous populations has a significant value, and is likely to remain sought-after for global research. A central concern of this book is that in none of the cases described above was there any clarity as to what the rights of the indigenous research population would be in the event that the research led to a scientific breakthrough and lucrative drugs in the developed world. Anticipated outcomes and advantages for the research organisations were not adequately mirrored in any form of assurance of commensurate benefit for DNA donors, leading to perceptions framed variously as unfairness, exploitation, neo-colonialism and even of biopiracy (Staples 2000). Admittedly, in some cases, the validity of claims of biopiracy was somewhat countered by the concern that the media's creativity in seeking sensation might have inappropriately raised the profile of the issue (Pomfret and Nelson 2000).

In seeking a balanced view of the issue it needs to be emphasized that research on indigenous peoples is not necessarily problematic. Press coverage is largely provided to those cases that are disputed, ignoring the many that proceed in good faith, with due attention to best practices and respecting the rights and sensitivities of the research populations. In some situations the researched indigenous populations appeared to have sufficiently comprehended the nature of the transaction, and to have cooperated voluntarily with researchers in the hopes of contributing towards public health. The Hagahai referred to above, for example, willingly agreed to donate their blood samples for the creation of cell lines, as well as to allow their patenting if appropriate, in a negotiated agreement that allowed for financial benefits in the event of a commercially successful product being discovered (Ibeji and Gane 1996). Furthermore, the Tristan da Cunha islanders negotiated a deal that would secure them future free asthma drugs in the event of success, even if the agreement failed to secure a share in the windfall profits generated downstream by the sale of the research discovery.

1.6 Concerns About the Fairness of Exchanges of Genetic Material

In each of the above cases, when these selected populations provided their samples for research purposes, the future and present value of their contributions would at that juncture have been speculative and therefore difficult to quantify. It is for that reason a legitimate and abiding concern that in each case the eventual value would normally be realized in developed countries, possibly years later, and far from the original exchange. An additional unease however, arising from a cursory glance at these cases, is the fact that the intrinsic value of samples is not limited to future commercial drug discoveries.

A significant form of consequence associated with genetic research and one far more difficult to quantify is the sheer importance and thus coupled value of possession of the samples, creating the potential in developed-world institutions for academic recognition and associated institutional advancement to follow. For example, the 400 samples of Arizona's Havasupai tribe formed the basis for 23 scholarly papers, articles and dissertations (Hendricks 2004), and the 833 samples taken from the Nuu-cha-nulth of Vancouver island formed the basis for the eventual promotion of the chief researcher Dr. Ward as the head of the Institute of Biological Anthropology at Oxford University (Malik 2005, p. 21). Similarly, Harvard University's joint project with Millennium Pharmaceuticals to gather tissue samples from a highly homogenous population of the Anhui region of China was brimming with this alternative form of research value, described by one researcher as being, "more valuable than gold" (Pomfret and Nelson 2000, p. 1).

This research value was summed up in the extract below from an article on the project in the Washington Post entitled, *An Isolated Region's Genetic Mother Lode*. It states that, "Harvard ultimately reaped millions of dollars in federal grants and private investment for the university because of its access to Anhui DNA, and Millennium was able to raise tens of millions of dollars from corporate investors" (ibid, p. 1). To conclude this point then, there are clearly many forms of value locked up in the human genetic resources of certain populations. These forms of value encompass not only the research value for global health, and the institutional value associated with the ability to license and share the information, but in addition its downstream commercial potential.

Where profits or other forms of value accrue to commercial entities resulting from research involving particular communities, it has been suggested that consideration should be given to the sharing of benefits with the broader community (Sheremeta et al. 2007, p. 161). Such a sharing in the event of a windfall discovery, or of direct profits flowing from the research, would appear at face value to be fair to all parties in the exchange. This begs the questions as to how the value should be acknowledged, or properly compensated for, in cases where profits are visible such as in the Tristan da Cunha case, as well as the less ostensibly commercial circumstances such as the Harvard University project referred to above.

Issues of fairness and justice in genomic research exchanges therefore range from the global level, involving developed and developing countries such as the Indonesian crisis, down to the local or community level where researchers engage with groups that are often isolated and vulnerable. Many commentators examining such genomic exchanges between researchers and indigenous or other developing-world rural communities have suggested that benefit sharing, as is legally required for non-human genetic resources under the Convention for Biological Diversity, is the mechanism most appropriate to establish fairness in this form of exchange (Merz et al. 2002; Hayden 2007; Simm 2005 and Schroeder and Pogge 2009). Thambisetti (2003) captures the alarm raised about genomic research in developing countries as, "a concern that the full benefit of the medical advances that may result from the decoding of the human genome may not be realised if the genes become subject to privately owned intellectual property and are exploited solely for profit" (Thambisetti 2002, p. 12).

Apprehensions such as these have drawn attention to the potential injustice, unfairness, or exploitation inherent in these types of transactions, and will receive further analysis below. The unfairness captured by the examples and highlighted by the supporters of benefit sharing referred to above could be summarised in the following progression of statements:

1. Genomic research places high value on accessing human DNA from particular and often vulnerable populations situated in the developing world.
2. Research is intended not only to advance scientific knowledge, but also to seek potentially valuable medical discoveries.
3. Significant value is received by research institutions not only due to possession of the samples enhancing research capacity, but also due to the potential for downstream commercial discoveries.
4. It is appropriate for reasons of fairness and equity to consider some form of acknowledgement, reciprocation, and in some cases even financial consideration for the value provided by the donor community.
5. Where no form of benefit sharing is contemplated in this exchange of human DNA, perceptions of injustice or exploitation are justly aroused.

To conclude this introductory chapter, the exchange of human biological resources from the developing to the developed world as part of global research has been broadly described, emphasising the increasing scope and importance of genomic science for global health. Concerns that there might be unfairness in such exchanges have been made not only by countries such as Indonesia, but also by rural and indigenous communities targeted for genetic research. These concerns of unfairness in the exchange motivate the overall purpose of this book, which will be achieved by answering three research questions, recapitulated as firstly, whether and in what circumstances benefit sharing is appropriate for human genetic research. Secondly, whether benefit sharing runs the risk of unduly inducing or coercing research participants and thirdly, how the particular concerns of indigenous peoples are addressed by existing research practice and guidelines.

Prior to examining the most significant objections that have been made against benefit sharing, Chap. 2 will examine the meaning of the word that is arguably relevant to all discussions relating to justice, that is to say *exploitation*. Thereafter Chap. 3 will analyse the first objection to benefit sharing, namely the assertion that human genetic resources should be freely shared as part of the *common heritage of humankind*. Chapter 4 examines the second objection to benefit sharing, namely the *altruism* argument, which states that humans should share their DNA with no expectation or return, in solidarity with humankind. Chapter 5 discusses a further argument against benefit sharing, namely that human donors have not *added value* to their samples, and are thus not deserving of any benefits. Chapter 6 prepares for the discussion of undue influence and coercion as the final objection to benefit sharing, by first examining the contribution of *justice and law* to understanding exploitation, and then discussing the legal mechanisms that evolved through the law of equity to counter exploitation in such exchanges. Chapter 7 analyses *undue influence* and *coercion* as the final and possibly most persistent objection to benefit sharing in genomic research, and Chap. 8 examines the particular risks and concerns experienced by *indigenous peoples* who form an important category of genomic research participants. The final chapter will draw out conclusions of this book.

References

Andanda, P. (2008). Human-tissue related inventions: Ownership and intellectual property rights in international collaborative research in developing countries. *Journal of Medical Ethics, 34*(3), 171–179.

Annas, G. (1990). Outrageous fortunes: Selling other people's cells. *Hastings Centre Report, 20*(6), 36–39.

Annas, G. (1999). Genetic privacy, there ought to be a law. *Texas Review of Law and Politics, 4*, 7–15.

Arnason, V. (2011). Database research: Public and private interests. *Cambridge Quarterly of Healthcare Ethics, 20*(4), 563–573.

Beauchamp, T. L. (2011). Informed consent: Its history, meaning and present challenges. *Cambridge Quarterly of Healthcare Ethics, 20*(4), 515–523.

Boggio, A. (2005). Charitable trust and human genetic databases: The way forward? *Genomics, Society and Policy, 1*, 41–49.

Brand, A., Brand, H., & Schulte, T. (2008). The impact of genetics and genomics on public health. *European Journal of Human Genetics, 16*, 5–16.

Brodwin, P. (2005). Bioethics in action and human population genetics research. *Culture, Medicine and Psychiatry, 29*, 145–178.

Caulfield, T., & Chapman, A. (2005). Human dignity as a criterion for science policy. *PLoS Medicine, 2*, 736–738.

Chennells, R., Haraseb, V., & Ngakaeaja, M. (2009). Speaking for the san: Challenges for representative institutions. In R. Wynberg, D. Schroeder, & R. Chennells (Eds.), *Indigenous Peoples, Consent and Benefit Sharing: Lessons from the San Hoodia Case* (pp. 165–189). Springer: London.

Chadwick, R. (2011) The communitarian turn: Myth or reality. *Cambridge Quarterly of Healthcare Ethics*, 546–553.

Christie, J. (1996). Whose property, whose rights? *Cultural Survival Quarterly, 20*, 34–36.

Cunningham, H. (1998). Colonial encounters in postcolonial contexts: Patenting indigenous DNA and the human genome diversity project. *Critique of Anthropology, 18*(2), 205–217.

Dalton, A. (2002). Tribe blasts 'exploitation' of blood samples. *Nature, 3*. http://www.natur.com/nature/journal/v420/n6912/full/420111a/html.

Dang, Z., & Lei, X. (1999). Chinese centre sues over study coverage. *Scientist, 270*, 1990–1992.

Darwin, E. (1794). *Zoonomia or the laws of organic life* (3rd ed.). (1801) London: J Johnson.

Dawkins, R. (1995). *River out of Eden.* Weidenfeld and Nicolson, London.

Dawkins, R. (2004). *The ancestor's tale: a pilgrimage to the dawn of life*. London: Phoenix, Orion Books Limited.

Dutfield, G. (2004). *Intellectual property, biogenetic resources and traditional knowledge.* London: Earthscan Publications Limited.

Friedlander, J. (1996). Genes, people and property: Furor erupts over genetic research on indigenous groups. *Cultural Survival Quarterly, 20*(2).

Fujimura, J. (2005, August). Postgenomic futures: Translations across the machine-nature border in systems biology. *New Genetics and Society, 24*(2).

Fujimura, J., & Rajagopalan, R. (2011). Different differences: The use of 'genetic ancestry' versus race in biomedical human genetic research. *Social Studies of Science, 41*, 5–30 [Online]. Retrieved from http://sss.sagepub.com/content/41/1/5.

Gibbons, S., Kaye, J., Smart, A., Heeney, C., & Parker, M. (2007). Governing genetic databases: Challenges facing research regulation and practice. *Journal of Law and Society, 34*(2), 165–167.

Greely, H. T. (2000). Iceland's plan for genomics research: Facts and implications. *Jurimetrics, 40*(2), 153–191.

Habeck, M. (2002, October 17). *Estonia jumps on the gene bank train.* The Scientist October 17th. www.biomedcentral.com/ness/20021017/08.

Hamilton, Cindy. (2001). The human genome diversity project and the new biological imperialism. *Santa Clara Law Review, 619*, 627–628.

Harmon, A. (2010, April 25). *Indian tribe wins fight to limit research of its DNA* (p. 21). New York Times. http://www.nytimes.com/2010/04/22/us/22dnaside.html.

Hayden, C. (2007). Taking as giving: Bioscience, exchange, and the politics of benefit sharing. *Social Studies of Science, 37*, 729–758.

Hendricks, L. (2004, April 16). *Havasupai tribe files $50 million lawsuit against ASU.* Arizona Daily Sun.

Herper, M. W. (2012). *The truly staggering cost of inventing new drugs.* Forbes. http://www.forbes.com/sites/matthewwherper/2012/02/10.

Hook, E. B. (1976). In A. E. H. Emery & J. R. Miller (Eds.), *Registers for the detection and prevention of genetic diseases* (pp. 9–29). Stratton, New York.

HUGO (2000). HUGO Ethics Committee. *Statement on benefit sharing*. Retrieved April 9, 2000, from http://www.hugointernational.org/Statement_on_Benefit_Sharing.htm.

Ibeji, I., & Gane, K. (1996). The hagahai patent controversy: In their own words. *Cultural Survival Quarterly Summer, 33*, 34–36.

ICESCR (1966) International Covenant on Economic, Social and Cultural Rights. *Adopted and opened for signature, ratification and accession by general assembly resolution 2200A (XXI)* of 16 December 1966, entry into force 3 January 1976, in accordance with article 27. www.unhchr.ch/html/menu3/b/a_cescr.

ILO. (1989). International Labour Organisation. Indigenous and tribal peoples. *Convention, 169.* http://www.ilo.org/dyn/normlex/en/f?p=NORMLEXPUB:12100:0::NO::P12100_ILO_CODE:C169.

International HapMap Consortium (2005). A haplotype map of the human genome. *Nature, 437*, 1299–1320. http://www.nature.com/nature/journal/v437/n7063/full/nature04226.html.

Jones, S. (2009). *The britannia guide to genetics.* London: Constable and Robinson.

Karlsen, J. R., Solbakk, J. H., & Holm, S. (2011). Ethical endgames: Broad consent for narrow interests; open consent for closed minds. *Cambridge Quarterly of Healthcare Ethics, 20*(4), 572–583.

Kegley, J. A. K. (2004). Challenges to informed consent. *European Molecular Biology Organisation, 5*(9), 832–836.

Khoury, M. J. (2004). The case for a global human genome epidemiology initiative. *Nature Genetics, 36*(10), 1027–1028.

Knoppers, B. M., Abdul-Rahman, M., & Bedard, K. (2007). Genomic databases and international collaboration. *Kings Law Journal, 18*, 291–312.

Knoppers, B. M., & Abdul-Rahman, M. H. (2008). Biobanks in the literature. In B. Elger & N. Biller-Andorno (Eds.), *Ethical Issues in governing biobanks*. London: Ashgate Publishing.

Kolata, G. (2003). *Using genetic tests, ashkenazi jews vanquish a disease*. New York Times. Retrieved February 18, 2003, from http://www.nytimes.com/2003/02/18/science/using-genetic-tests-ashkenazi-jews-vanquish-a-disease.html?pagewanted=allandsrc=pm.

Kowal, E. (2012). Genetic research in indigenous health: Significant progress, substantial challenges. *Medical Journal of Australia, 197*(1).

Laird, S. A., & Wynberg, R. P. (2002) Institutional policies for biodiversity research: Setting standards for conduct, prior informed consent, and benefit sharing. In S. A. Laird (Ed.), *Biodiversity and traditional knowledge: Equitable partnerships in practice*. London: Earthscan.

Lehrman, S. (1996) US drops patent claim to Hagahai cell line. *Nature, 384*, 500.

Levene, S. (2004). *Genetic screening in the ashkenazi jewish population—Is there a case for more testing? Guy's Hospital report for National Screening Committee*, Retrieved June 3, 2004, from http://www.screening.nhs.uk/policydb_download.php?doc=35.

Lock, M. (1997). Decentering the natural body: Making a difference matter. *Configurations, 5*(2), 267–292.

MacIntosh, C. (2005). Indigenous self-determination and research on human genetic material. *Health Law Journal, 13*, 213–250.

Macklin, R. (2004). *Double standards in medical research in developing countries*. Cambridge: Cambridge University Press.

Malik, P. (2005). Biopiracy. *Canadian Journal of Cardiology, 21*(1), 21.

McGregor, J. L. (2007). Population genomics and research ethics with socially identifiable groups. *Journal of Law, Medicine and Ethics, 35*, 356–370.

Mehlman, M. J., & Botkin, J. R. (1998). *Access to the genome: The challenge to equality*. Washington DC: Georgetown University Press.

Merz, J., Magnus, D., Mildred, K., & Caplan, A. (2002). Protecting subjects: Interests in genetics research. *American Journal of Human Genetics, 70*(4), 965–71.

NIH. (2008). Three sequencing companies join 1000 genomes project. National Institutes of Health News. Retrieved January 23, 2008, from http://www.nih.gov/news/health/jun2008/nhgri-11.htm.

NIH. (2012). National Institutes of Health Funding. *NIH funding opportunities and notices*. Retrieved August 22, 2012, from http://grants.nih.gov/grants/guide/rfa-files/RFA-RM-12-006.html.

O'Neill, O. (2004). Informed consent and public health. *Philosophical Transactions of the Royal Society: Biological Sciences, 359*(1447), 1133–1136.

Parry, J. (1986). The gift, the Indian gift, and the "Indian Gift". *Man (n.s), 21*(3), 453–473.

Pellekaan, Van der Hoorst, S. (2002). Genetic research: What does this mean for indigenous australian communities? *Australian Aboriginal Studies*, 65–75. http://www.questia.com/library/journal/1G1-72609271/genetic-research-what-does-this-mean-for-indigenous.

Pharma. (2012). *The big pharma recession report*. www.pharma-iq.com/downloadContent.cfm.

Pogge, T. W. (2008). What is global justice? *Revista de Economía Institucional, 10*(19), Second Semester 2008.

Pomfret, J., & Nelson, D. (2000). In rural China, a genetic mother lode. *The Body Hunters*, Article IV, The Washington Post. Retrieved December 20, 2000, from http://www.washingtonpost.com/wp-dyn/content/article/2008/10/01/AR2008100101158.html.

Price, M. (2012). *Worth looking into*. Sunday Times. www.sundaytimes.co.za.

Rajan, K. (2006). *Biocapital: The constitution of postgenomic life*. Durham: Duke University Press.

Reddy, D. S. (2007). Good gifts for the common good: Blood and bioethics in the market of genetic research. *Cultural Anthropology, 22*(3), 429–432.

Ridley, M. (1999). *Genome: The autobiography of a species*. London: Fourth Estate.

Richards, M. (2001). How distinctive is genetic information. *Studies in History and Philosophy of Science Part C: Biological and Biomedical Sciences, 32*(4), 663–687.
Roberts, L. (1991). Fights erupts over DNA fingerprinting. *Science, 254*, 1721–1750.
Rose, N. (2004). Developing our links with China; sociology and BIOS. *Sociology Research News, 3*(1), 6.
Rowe, P. (1996). Lessons about NIDDM from the Pima Indians. *Lancet, 347*, 1320.
Salopek, P. (1997). Genes offer sampling of hope and fear. *Chicago Tribune 28*, part 1. http://www.chicagotribune.com/news/local/chi-060826salopekgene2-story,0,6138822.story.
Saprai, P. (2009). The principle against self-enslavement in contract law. *Journal of Contract Law, 26*(1) 22–44.
Schroeder, D., & Pogge, T. (2009). Justice and the convention on biological diversity. *Ethics and International Affairs, 23*, 265–278.
Sedyaningsih, E. R., Siti, I., Soendoro, T., & Sitit, S. (2008). Towards mutual trust, transparency and equity in virus sharing mechanism: The avian influenza case of indonesia. *Annals Academy of Medicine. 37*(6), 262–268.
Sheremeta, L., & Knoppers, B. M. (2004). Beyond the rhetoric: Population genetics and benefit sharing. *Health Law Journal, 11*(2), 89–117.
Sheremeta, L., & Knoppers, B. M. (2007). Beyond the rhetoric: Population genetics and benefit sharing. In B. Phillips & C. B. Onwuekwe (Eds.), *Accessing and Sharing the Benefits of the Genomics Revolution* (pp. 157–182). Springer.
Simm, K. (2005). Benefit sharing: An enquiry regarding the meaning and limits of the concept in human genetic research. *Genomics, Society and Policy, 1*(2), 29–40.
Simm, K. (2011). The concepts of common good and public interest: From plato to biobanking. *Cambridge Quarterly of Healthcare Ethics, 20*(4), 554–562.
Slabbert, M., & Pepper, MS. (2010). A room of our own? Legal lacunae regarding Genomic Sovereignty in South Africa. *Tydskrif vir Hedendaagse Romeinse Hollandse Reg, 73.*
Soodyal, H., Nebel, A., Morar, B., Jenkins, T. (2003). Genealogy and genes: Tracing the founding fathers of Tristan da Cunha. *European Journal of Human Genetics, 11*, 705–709.
Staples, S. (2000). Human resource: Newfoundland's 300-year old genetic legacy has triggered gold rush. *Report on Business, 17*(3), 117–120.
Statistics Canada. (2002). *Estimates of total research expenditures on research and development in the health fields in Canada, 1988 to 2001.* http://www.statcan.gc.ca/pub/89f0115x/89f0115x2013001-eng.htm.
Tauli-Corpuz, V. (2003). *Biodiversity, traditional knowledge, and rights of indigenous peoples* (p. 33). Penang: Third World Network. http://www.twnside.org.sg/title2/IPR/pdf/ipr05.pdf.
Thambisetti, S. (2002). Patents and human genome research in developing countries. *Journal of World Intellectual Property, 5*(5), 683–723.
The Economist. (1998, February 21–27). *The pharmaceutical industry survey.*
The Economist. (2008, July 24). *All together now: Liberalisation and the quest for scale are pushing generic drugs firms to merge* (pp. 57–58), New York.
Tierney, P. (2000). *Darkness in El Dorado: How scientists and journalists devastated the Amazon.* New York: Norton. 2000.
Tokar, B. (2002). *Your genes, owned by others: Mining humanity.* Third World Network. http://www.progress.org/toka.htm.
UNDRIP. (2007). United Nations Declaration on the Rights of Indigenous Peoples. www.un.org/esa/socdev/unpfii/documents/DRIPS.
UNESCO. (2005). Universal Declaration on Bioethics and Human Rights (1997). Human Rights. United Nations Educational Scientific and Cultural Organization, Article 2 (f). http://www.unesco.org/new/en/social-and-human-sciences/themes/bioethics/human-genome-and-human-rights/.
Upshur, R., Lavery J. O., Tindana, P. (2007). Taking tissue seriously means taking communities seriously. *BMC Medical Ethics, 8*(11), 3–6.
Uranga, A. M. (2005). *Outstanding legal and ethical issues on biobanks: An overview of the regulations of member states of the euro bio-bank project.* Madred: Instituto de Salud Carlos III.

Waitangi, T. (2011). *Ko aotearoa tenei: A report into claims concerning new zealand law and policy affecting maori culture and identity*, report 262 [Online]. Retrieved from http://www.justice.govt.nz/tribunals/waitangi-tribunal/news/wai-262-ko-aotearoa-tenei-report-on-the-wai-262-claim-released.

Wang, Y. (2011). *Group protection in human population genetic research in developing countries: The people's republic of China as an example*. Unpublished Ph.D. book, University of Glasgow. http://theses.gla.ac.uk/3005/1/2011WangYuephd.pdf.

Weijer, C. (1999). Protecting communities in research: Philosophical and pragmatic challenges. *Cambridge Quarterly for Healthcare Ethics, 8*, 503–513.

Weijer, C., Golds, G., & Emanuel, E. J. (1999). Protecting communities in research: current guidelines and limits of extrapolation. *Nature Genetics, 23*, 275–280.

Weijer, C., & Miller, P. B. (2004). Protecting communities in pharmacogenetic and pharmacogenomic research. *The Pharmacogenomics Journal, 4*, 6–9.

WHO. (2002). World Health Organisation Report of the Advisory Committee on Health Research, *Genomics and world health*. http://whqlibdoc.who.int/hq/2002/a74580.pdf.

Williams, G. (2005). Bioethics and large-scale biobanking: Individualistic ethics and collective projects. Genomics. *Society and Policy, 1*(2), 50–66.

Williams, G., & Schroeder, D. (2004). Human genetic banking: Altruism, benefit and consent. *New Genetics and Society, 23*(1), 89–103.

Wilke, M. (2012). The *world health organizations pandemic influenza preparedness framework for the sharing of influenza viruses and access to vaccines and other benefits. ABS in the health arena*. Unpublished. http://www.feu.unibremen.de/cms/fileadmin/dokumente/Workshop/Abstracts/Marie_Wilke_WHO_pandemic_influenza_preparedness_framework.pdf.

Winickoff, D. (2003). Governing population genomics: Law, bioethics and biopolitics in three case studies. *Jurimetrics, 43*(2), 187–227.

Zamel, N. (1995). In search of the genes of asthma on the island of tristan da cunha. *Canadian Respiratory Journal*. 2(1), 18–22.

Zamel, N., McClean, P. A., Sandell, P. R. (1996). Asthma on tristan da cunha. *American Journal of Respiratory Critical Care Medicine*, *153*(6), 1902–1906.

Zhou, Q. (2006). The legal status of human gene and genetic information (Lun Renlei Jiyin ji Jiyinxinxi de Falvdiwei). *Science-Technology and Law (in Chinese), 63*, 113–117.

Chapter 2
Exploitation

Abstract The aspiration to ensure access to human DNA without exploitation of research communities being a central theme of the book, this chapter explores meanings of the word *exploitation* from its everyday usage to its various interpretations as a philosophical concept. The most significant contributions from the literature are traversed, such as *advantageous use of someone*, or *interaction stained by wrong*, or *gaining advantage at the expense of another*. Legal approaches to exploitation are introduced, with substantive and procedural fairness as themes that are revisited throughout the book. Likewise, non-legal approaches are introduced such as Wertheimer's *taking unfair advantage,* focus on *consent* as an indicator of fairness, and the *Principle of Permissable Exploitation*. Finally Mayer's three forms of exploitation are chosen as providing clear guidance for the questions being examined. The three forms are firstly where exploiters do not benefit their victims at all (the free rider), secondly where they do not benefit them sufficiently (an unfair exchange), and thirdly where they do not benefit them authentically (an illegal or highly immoral transaction, e.g. the provision of heroin at reasonable prices). The chapter closes by explaining how Mayer's last two classes will be used as yardsticks in subjecting the exchange of human DNA to scrutiny.

Keywords Exploitation · Legal exploitation · Non-legal exploitation · Principle of permissible exploitation · Fair exchange

As stated by Ruth Macklin in her study of double standards in medical research, concerns about justice in international research have, "focused mostly on worries about exploitation of human research participants or indeed of entire populations in the developing countries" (Macklin 2004, p. 68). Many authors have claimed that research participants in the developing world are exploited, one of the key assertions being that they do not receive a fair share of the benefits (Brody 2001, p. 2857). One of the primary aims of the CBD in its legal governance of the utilisation of non-human genetic resources was to prevent the perceived exploitation of developed countries. The term *biopiracy,* which has been coined to describe

R.S. Chennells, *Equitable Access to Human Biological Resources in Developing Countries*, DOI 10.1007/978-3-319-19725-8_2

inherently unfair transactions involving genetic resources and is discussed in the next chapter, is predicated upon the concept of exploitation of individuals and groups, the prevention of which is the purpose of ethical guidelines for biomedical research.

This book examines the exchange of human biological resources across national borders, and seeks in particular to determine how scientific and medical research should utilize human biological resources without exploiting individuals, populations and even countries. It is therefore essential at the outset to explore and establish the meaning of this term *exploitation*, used so freely whenever claims of injustice or lack of fairness are made. Starting with everyday meanings of the term, attempts to understand the concept by various authors will be traversed followed by a discussion of how exploitation is described and dealt with both in the legal and non-legal context. The three forms of exploitation proposed by Mayer (2007) will then be described and assessed, seeking to provide a practical platform from which exchanges of human genetic resources, and arguments both for and against benefit sharing, can be examined.

2.1 Everyday Use of the Word Exploitation

Exploitation is at once an intuitively simple but also an ethically diffuse and unclear concept. The term is central to certain Marxist, neoclassical and liberal theories of economics, each of which contains firm assumptions about justice as well as a strong emotive component about the form of exploitation to be avoided. Activists such as Vandana Shiva utilised words such as piracy or biopiracy to denote a particular form of exploitation of developing countries by Western nations, whilst Che Guevara referred to the plunder of the poor in a speech to the United Nations (Guevara 1964). The word implies one-sided transactions that are perceived in some way as wrongful and to be avoided, and is strongly associated with concepts such as vulnerability, dignity and fairness in research that are deeply implicated in the field of bioethics. Authors differ widely on its precise meaning, one for example calling exploitation a "moral red herring" which nothing but a context-specific or case-by-case analysis can adequately resolve (Carse and Little 2008, p. 207), whilst others use utilitarian or Kantian ethics to arrive at its meaning.

Many of the definitions used are less than clear, such as Arnason's, "exploitation is interaction that is stained by wrong" (Arnason 2006, p. 162). Dictionary definitions separate two primary forms of the word, the first being an unfair treatment of someone, or the use of a situation in a way that is wrong in order to get some benefit for oneself, and the second one being the process of making use of something so that you gain as much as possible from it.[1] An example of the first

[1] [Online] Available at http://www.macmillandictionary.com.

form is the exploitation of illegal immigrant labour, and of the second is the commercial or economic or industrial exploitation of resources such as mining or timber extraction. Other uses of the word denote activities that are not necessarily ethically wrong. For example, the exploiting of one's talent, of natural resources, or even of the slowness of a business opponent all seem to be morally permissible, whilst exploiting the vulnerability or ignorance of another person in a business deal is generally regarded as less so.

2.2 Attempts at Understanding the Concept

Hawkins and Emanuel (2008) propose that the proper meaning of the word *exploitation* requires that the *advantageous use* of something or someone by another is required to be in some sense unfair or "morally problematic" (ibid, p. 44). They note that there is general consensus around three structural issues (ibid, p. 46). Firstly, regarding voluntariness, there is consensus that exploitation can take place voluntarily between willing parties and yet still be morally problematic. In other words, it is not necessary that one party be coerced or forced for exploitation to have taken place, and the wrongfulness of the transaction, if any, might lie in the distributive unfairness of the transaction, or in the deep vulnerability of the exploitee, under particular circumstances (ibid, p. 47). An example of this would be an employer taking on child labour from an extremely disadvantaged community and paying excessively low wages. Secondly, regarding the presence of harm, they believe that there is consensus that exploitation can but does not *need* to be harmful for the exploitee, for one can be wronged without being harmed.

Admitting that there is a loose sense of the word *harm* in which harm is just a synonym for wrong, they suggest that substantive criteria and reasons are required to determine the exact and contextual meaning of the word (ibid, p. 48). One example of this latter form of exploitation would presumably be secretly taking a photograph of an indigenous villager, with the intention of selling the photograph for a large amount of money. On the one hand the villager has not suffered any overt form of harm. On the other hand, she has not had a chance to object.

The famous case of Henrietta Lack, who died before the unauthorised utilisation of her tissue created highly profitable cell lines, was another example of no direct harm for the exploitee, whilst others subsequently denounced her treatment as being exploitative (Skloot 2010). Hawkins suggests that this form of exploitation, in which wrongs occur in the absence of direct harm, are regarded by many authors as being the most troubling of all (Hawkins et al. 2008, p. 48). Thirdly, regarding prohibition of exploitation, Hawkins notes a consensus that despite the fact that a practice may be exploitative, it does not automatically follow that it ought to be prohibited. Not all morally troubling acts can be effectively regulated by the state or the law. An example of this form, according to Hawkins, would be clinical trials in developing world countries, where people are desperately poor and the health system virtually non-existent (ibid, p. 48).

Carling adds another approach to determining the existence or otherwise of exploitation in an exchange, by asking two questions. Firstly, does one individual or group gain an advantage at the expense of another? Secondly he asks a normative question, namely is this advantage unjustly obtained? (Carling 1998, p. 220) The second question is more complex, and usually involves an analysis of what is meant by gain at the expense of another. One school of thought holds that such transactions are exploitative and wrong when and because they are coercive of the weaker party (Buchanan 1979; Schwartz 1995), whilst others hold that the wrongfulness lies in its degrading of the victims; when they are treated simply as means, not as ends (Wood 1995).

This book examines concerns about exploitation between parties where one party is frequently more vulnerable, and where various power and other differentials exist, leading to the possibility of injustice in the exchange. This question is therefore examined at greater depth in Chap. 6, which explores notions of justice, as well as contractual responses to the different forms of exploitation, which can take place between two transacting parties.

2.3 Legal Approaches to Exploitation

The law on contract, discussed further in Chap. 6 below, has evolved as a framework to regulate voluntary exchange transactions, during which parties of different power and resources negotiate or bargain and compete for advantage. Bargaining is according to Fried an enterprise involving risks for both parties, which "lacks both written rules and the scrutiny of an umpire [...] Ethical constraints may therefore serve as a surrogate for procedural rules" (Fried 1978, p. 109). Justice during such exchanges is a relational concept, which balances one person's interests with others. Bigwood (2003) describes the evolution of the law against contractual exploitation, which can take place during a voluntary exchange between two (or more) parties. He explains that the underlying principle supporting such evolution, which was influenced by the law of equity and invariably involves protecting the weaker party, is captured in the sentence "other things being equal, we believe that a party should be excused from performing a contract that resulted from the other person's exploitation" (Bigwood 2003, p. 1).

Judges have been required to decide and achieve legitimacy in countless exploitation cases over the centuries, without the same metaphysical luxuries enjoyed by philosophers and other commentators on justice who do not have their theories tested in practice (ibid, p. 8). In other words the law provides an account of how those judges who were assigned the responsibility to decide on legality, which included the questions of fairness and justice in particular transactions, determined the cases placed before them. Bigwood describes how exploitation in contract law is conceptualised in different ways, some examining the *mode of conduct* of the exploiter in more detail, which focuses on the conduct of the parties or the *procedural fairness*, whilst others examine the *quality of the outcome* of a transaction, also termed *substantive fairness*. In making the point that substantive unfairness in the form of an outcome that is bad for one party is not essential for exploitation to exist, he emphasises that to exploit someone in contract formation is to wrong her or treat her unjustly, regardless of how much she may lose, or her exploiter may gain, materially from the event. In other words, he holds that whilst exchange imbalance may vitally assist in the determination of an exploitation claim, the demerits of the bargain struck are not themselves objects of evaluation, and are not essential to the existence of exploitation (ibid, p. 6). As discussed below, exploitation can take place whilst improving the lot of the exploited person.

2.4 Non-legal Approaches to Exploitation

Returning to the basic meaning of the word exploitation, Wertheimer explains that exploitation can be used in a non-moral or non-derisive sense when it means simply to use or take advantage of, and that it is necessary to determine precisely the grounds upon which exploitative transactions may be prohibited (Wertheimer 2008, p. 66). Using *taking unfair advantage* as a yardstick, he states that exploitation can be determined in one of two ways. Firstly, some dimension of the outcome of the exploitative act needs to be assessed, which consists of an analysis of (1) the benefit to A and (2) the effect on B. The second dimension or test of unfair advantage examines whether there is some defect in the process whereby the (unfair) outcome is reached.

The process might be exploitative for example, where A unduly coerces B into giving consent, or deceives B, or fails to provide B with relevant information, or B is unable to understand the information adequately, all which are components of the informed consent process. In the latter case, where B is unable to understand the information, the unfairness if any would presumably reside either in A having known of or being able to predict B's lack of understanding, or alternatively where B indicated an inability to understand the information. Exploitation in the form of wrongful coercion might include a seductive offer that for some reason distorts the judgment of B, or which induces a lack of voluntariness in response, both modes of conduct which deserve and will receive more detailed analysis in Chap. 6 below.

It is noted that the above analysis of exploitation as *taking advantage* does not deal with the situation where there is a moral defect in the process rather than the outcome, for example where A unduly coerces B into consenting to a transaction, but where the outcome is in fact advantageous to B. This would entail some version of benign or patronising dominance of another, for B's own good. An example might be a company forcing a tribe to sign an agreement to provide traditional knowledge about a plant for commercial purposes, by withholding certain information deemed to be likely to cause confusion and prevent the contract, leading to an eventual commercially beneficial outcome, which helps the community significantly and does not infringe any of their cultural values. Still a moral assessment of such a process and outcome—a disrespectful process leading to a good outcome—might well be acceptable to a utilitarian, but is likely to be unacceptable to a Kantian analyst.

Where both parties benefit from a transaction, no matter how exploitative, Wertheimer describes such a mutually advantageous transaction as being *Pareto superior* (Wertheimer 1996, p. 14). He acknowledges that finding an acceptable standard of exploitation is no simple matter, and suggests,

> We appear to be stuck. Although I cannot produce a non-problematic theory of fair transactions, I remain convinced that some mutually advantageous transactions are quite unfair and exploitative. I am reluctant to mimic Justice Potter Steward's view of pornography with respect to exploitation, namely "I shall not attempt to further define pornography, but I know it when I see it". (Wertheimer 2008, p. 73)

Consent is subjected to rigorous examination in Wertheimer's analysis, defects in consent being critical to an enquiry into the fairness of a transaction. The importance of such an enquiry resonates strongly with the law of equity discussed in Chap. 6, where the validity of a contract is determined entirely by the procedure surrounding and quality of the consent. Proper or authentic consent is an essential component of procedural justice that is required to defeat exploitation. Support for this focus on consent comes from Carse and Little, who avoid the red herring surrounding the concept of exploitation by conducting what they term an *essentially contextualist* enquiry into all the norms, facts and options surrounding the transaction (Carse and Little 2008, p. 207). They contend that morally asymmetrical exchanges take place when consent is for some reason, vulnerability or otherwise, wrongfully obtained (ibid, pp. 211–214).[2]

A final and useful aspect of Wertheimer's analysis is his Principle of Permissible Exploitation, which he refers to as PPE (Wertheimer 2008, p. 83). PPE holds that we should allow a transaction whenever it would be better for all parties to the transaction and worse for nobody else. That we should not interfere where mutually and consensually the parties agree to an exchange that is in some way disadvantageous to one. This is a consequentialist and practical approach to deal with lesser forms of imperfect transactions that are acceptable to the parties. Finally Wertheimer admits that PPE arouses concerns, and suggests that it is a plausible principle of *non-ideal* moral theory.

As opposed to Rawls, who restricted himself to developing an ideal theory of a just society, *non-ideal moral theory* aims to provide principles whereby individuals should act under the unjust and non-ideal conditions that pertain (ibid, p. 84). As shown above, some genetic research in developing countries is carried out in non-ideal circumstances, for example a lack of local research ethics committees (RECs), or an extremely vulnerable population, that are to some degree tainted with perceptions of unfairness. The research projects described above all involved

[2]Carse and Little (2008) refer to marketplace norms, which denote a transaction as morally permissible as long as both parties are competent, fully informed and sufficiently instrumentally rational, and there is no deception, fraud, manipulation, threat, coercion or force.

the taking of genetic samples from rural and often illiterate populations from the developing world, such as Tristan da Cunha and China as well as indigenous tribes such as the Hagahai, the Havasupai, the San, and the Australian Aborigines. Such cases call for reliable assurance that the research participants, obviously vulnerable in so many ways, are treated fairly.

Cori Hayden in an analysis of the complexities of the exchange of human DNA for genetic research, proposed an even simpler formulation in relation to consent for research, namely that research should "include people well" (Hayden 2007, p. 746). In a more just world, the present disparity between the power and fortunes of nations and peoples that led to exploitation of the weak by the strong would simply not exist. We are however, saddled with the fact that the concept of exploitation is central in the non-ideal environment that does exist, and remains difficult to constrain within a single definition.

2.5 Mayer's Three Forms of Exploitation

Exploitation remains a puzzling concept, complicated by the fact that under the principles of PPE, it is regarded as being sometimes permissible. We are left with the unanswered question, namely when and under what circumstances it can be identified or prohibited? Bearing in mind Wertheimer's admission on the futility of finding a unified theory of fair transactions, as well as the numerous scenarios within which genetic research is conducted, this book needs a clear and practical definition of exploitation that can be applied to evaluate the fairness of genomic exchange transactions involving humans. Robert Mayer (2007) addressed this question in a paper entitled, *What's Wrong with Exploitation?* in which he first singled out two terms, namely the "failure of reciprocity," and "wrongful gain" as being at the heart of what the distinction tries to achieve (ibid, p. 139). Wrongful gain, according to Mayer is the broader term, which is not always called exploitation. A thief who exploits one's carelessness and enters an open window to steal has broken the law and is called a thief, not an exploiter. Exploiters are those who gain illegitimately, or take advantage in a way that is wrong, not necessarily illegal, at another's expense, failing to reciprocate or benefit the disadvantaged party as fairness requires.

Under the broader criterion of wrongful gain therefore, a thief, an extortionist and a blackmailer are guilty of named offences, which whilst including elements of exploitation, do not benefit from being defined as such. The law defines these acts as criminal acts. The term exploitation would according to Mayer more appropriately describe the actions of free riders, drug pushers, slave masters, monopolists, usurers, sweat shop employers and so forth, who exact advantage in a morally dubious manner without necessarily breaking the law (ibid, p. 139). In other words, whilst many criminal (theft) or illegal (blackmail) acts contain an element of exploitation, the proper meaning of the word should reside outside of such particular forms of wrongdoing.

Using the word fairness as the touchstone, Mayer identifies three classes of exploitation, namely the first class, where the exploiters do not benefit their victims at all, the second class, where they do not benefit them sufficiently, and the third class, where they do not benefit them authentically (ibid, p. 143). In returning to the question of what fairness requires, Mayer reminds us firstly, that the loss suffered by an exploited party is always relative, rather than absolute (ibid, p. 141), and secondly, that "exploitation [...] is a thoroughly political concept because contestable ideas about what fairness requires determine whether taking unfair advantage is recognized or not" (ibid, p. 144).

Returning to Mayer's classifications, class 1 exploiters fail to benefit the disadvantaged party at all, the free rider being the classic example. A free rider is a person who enjoys the donuts purchased for all at the office, but fails to provide her financial contribution when the hat is passed around for contributions. Or one who enjoys the entire performance of a street performer, then walks away without making payment. Notably, this first class of dealing does not take place in the context of an explicit exchange, and is therefore not relevant for this book which deals with the exchange of human genetic resources.

Class 2 exploiters, who fail to benefit their victims sufficiently, are according to Mayer always involved in an exchange, in which the exploited party gains in relation to the *status quo ante*, but loses from the standpoint of fairness. She gains, but not sufficiently. Fairness in this sense is determined by an enquiry into a just value for a particular good, namely whether there was fairness in exchange. An example would be where a rich man fails to pay a starving man's asking price for a box of apples, knowing that the poor man has few other options and will accede to the low offered price. The power differential combined with the vulnerability of the weaker party is usually apparent to an objective observer, and arouses feelings of unfairness.

Finally, class 3 exploiters fail to benefit the exploited party *in an authentic manner*. This type of transaction also involves an exchange between two parties, but one that is regarded as *inauthentic* in that it should not, according to publicly accepted rules or norms, take place at all. An example of class 3 exploitation suggested by Mayer is the drug-pusher who sells to an addict. The addict chooses to purchase the drugs and the price is reasonable, but society would regard the drug-pusher as being exploitative, the transaction being harmful and wrongful, and one that should be prohibited (Mayer 2007, p. 145). This book proceeds from the assumption that the exchange of human genetic resources for research purposes *is* indeed an authentic, or legitimate practice, and for that reason Mayer's class two form of exploitation is primarily relevant to the enquiry into the fairness of genomic research exchanges.

How then to apply Mayer's second-class test? After exploring the slippery nature of the term, it can be concluded that the second class of exploitation identified by Mayer, which involves reciprocal exchanges where fairness is being questioned, where the exploiter fails to benefit their victims sufficiently, is entirely germane for the first part of this study, and will provide a useful framework for analysis of exchanges in genomic research. All of the complaints and concerns relating to bioprospecting discussed above had at their core accusations of exploitation, in the form of transactions, which are legally unassailable, but are redolent of short-changing, cheating, or some form of cheapness.

A closer look at the Tristan da Cunha case informs a simple assessment of fairness. In the initial transaction, the islanders gave consent to the company for the use of their DNA samples in exchange for a guarantee of access to a cure for asthma in the event of a possible future cure. When the company sold the rights to their subsequently discovered research lead for $70 million to another company, none of this benefit was shared with those who had provided the research samples. Application of Mayer's second class of exploitation test would simply enquire whether the first company, *Sequana Therapeutics,* benefited the islanders sufficiently.

Implementation of this test would admittedly require access to credible norms and standards in order to evaluate the fairness, or the equivalence of the exchange. Opponents of this test might therefore argue that no such generally acceptable criteria exist, or alternatively that the wide variety of circumstances surrounding such exchange transactions would make the application of such a test inoperable. Moreover, they might suggest that exploitation in an exchange can be determined by the application of a different set of criteria. A further objection might suggest that an examination of the actual process of engagement, namely the dealings, interactions and motives of the parties, would contribute more usefully towards an assessment of exploitation. Whilst a clear standard against which the outcome of an exchange is not readily available, it is suggested that an examination of the facts of a case on the basis of generally accepted standards does provide an observer with the ability to formulate an opinion as to whether the weaker party, the islanders in this case, were benefited insufficiently, and thus exploited.

2.6 Conclusions

During the discussion on the legal meanings of exploitation above, it was explained that the law of equity favoured an altogether different approach to exploitation, namely one, which emphasised the mode of conduct or unfair manner, employed by the exploiter, rather than the outcome or negative result suffered by the exploitee. This latter and more procedurally structured enquiry into the fairness of transactions will be further discussed in Chaps. 6 and 7 below. As every transaction comprises procedures, motives and relationships on the one hand, as well as substance and material outcomes on the other, both of these approaches are necessary and need to be integrated during an analysis of exploitation in the research exchange of human DNA.

To conclude this discussion, avoidance of Mayer's second and third forms of exploitation in the exchange of human genetic resources based upon considerations of fairness requires ensuring that the providers of genetic resources should be benefited sufficiently, and authentically, respectively. The former leads back to the proposal made by a range of commentators referred to in the first chapter, namely that benefit sharing is necessary in order to satisfy justice and avoid exploitation in research exchanges involving human DNA. This proposal for benefit sharing, the subject of the first research question posed in this book, now needs to address four objections in order to secure legitimacy. The first of the objections to be addressed below is the suggestion that human tissue should be shared freely as part of the common heritage of humankind.

References

Arnason, G., & van Niekerk, A. (2006). Undue fear of inducements in research in developing countries. *Cambridge Quarterly of Healthcare Ethics, 18*(2), 122–129.

Bigwood, R. (2003). *Exploitative contracts*. Oxford: Oxford University Press.

Brody, H. (2001). *The other side of eden: Hunter-gatherers, farmers, and the shaping of the world*. Canada: North Point Press.

Buchanan, A. (1979). Exploitation, alienation and injustice. *Canadian Journal of Philosophy, 9*, 138.

Carling, A. (1998). Exploitation. In *Encyclopedia of applied ethics* (Vol. 2). San Diego: Academic Press.

Carse, A. L., & Little, M. O. (2008). Exploitation and the Enterprise of Medical Research. In J. S. Hawkins & E. J. Emanuel (Eds.), *Exploitation and developing countries: The ethics of clinical research* (pp. 206–245). Princeton and Oxford: Princeton University Press.

Fried, C. (1978). *Right and wrong*. MA, Cambridge: Harvard University Press.

Guevara, C (1964). Speech delivered on March 25, 1964 at the plenary session of the United Nations Conference on Trade and Development (UNCTAD) [Online] Available in http://www.rcgfrfi.easynet.co.uk/ww/guevara/.

Hawkins, J., & Emanuel, E. J. (2008). *Exploitation and developing countries: The ethics of clinical research*. New Jersey: Princeton University Press.

Hayden, C. (2007). Taking as giving: Bioscience, exchange, and the politics of benefit sharing. *Social Studies of Science, 37*, 729–758.

Macklin, R. (2004). *Double standards in medical research in developing countries*. Cambridge: Cambridge University Press.

Mayer, R. (2007). What's wrong with exploitation? *Journal of Applied Philosophy*, 24(2).

Schwartz, J. (1995). What's wrong with exploitation? *Noûs, 29*, 159–164.

Skloot, R. (2010). *The immortal life of henrietta lacks*. New York: Crown Publishers.

Wertheimer, A. (1996). *Exploitation*. Princeton: Princeton University Press.

Wertheimer, A. (2008). Exploitation in Clinical Research. In J. S. Hawkins & E. J. Emanuel (Eds.), *Exploitation and developing countries* (pp. 63–104). Oxford: Princeton University Press.

Wood, A. (1995). Exploitation. *Social Philosophy and Policy, 12*, 150–151.

Chapter 3
Common Heritage of Humankind

Abstract This chapter introduces the first of four arguments against benefit sharing for the research utilization of human DNA, namely that it should be freely shared as part of what is commonly termed the *common heritage of humankind.* The trajectory of the common heritage concept is traced from John Locke to Hume and others that describe the type of resources that should not be subject to private dominion, finding expression in international laws such as the Law of the Sea and the Moon Treaty. The Convention for Biological Diversity (CBD) of 1992, which was negotiated in response to the perceived 'bioprospecting' of genetic resources by wealthy nations, provided a binding legal framework for their international exchange, and requiring benefit sharing to bring about fairness in all exchanges of biological resources. This chapter examines the governance of human DNA, which was first included in and then excluded from the CBD ambit, and was subsequently treated with increasing consensus between the Human Genome Organisation (HUGO), UNESCO and other institutions. The chapter ends with a comparison of bioprospecting of non-human and human genetic resources, concluding firstly that they are practically indistinguishable from one another in their research use, and secondly that neither can be said to form part of the common heritage of humankind.

Keywords Common heritage of humankind · Benefit sharing · Genetic resources · Fairness in exchange · Human genetic resources · Bioprospecting · Biopiracy

The case has been made in a broad and general manner for benefit sharing to bring about fairness in genetic research transactions. However, commentators opposed to benefit sharing in genetic research have raised a range of objections to this suggestion, the first of which is the assertion that human DNA should be regarded as part of the *common heritage of humankind.*

The common heritage of humankind assertion holds that there can be no question of exploitation or of unfairness in the exchange of human DNA for the simple reason that it forms part of the common heritage of humankind. As discussed in the first chapter, an estimated 99.8 % of human genetic information is non-differential, in other words is shared between all humans. This supports the argument that it should

R.S. Chennells, *Equitable Access to Human Biological Resources in Developing Countries*, DOI 10.1007/978-3-319-19725-8_3

be available for the benefit of all humankind (Wang 2011, p. 8). The argument continues to hold that human DNA is shared by all, such as the air we breathe, therefore no individuals have the right to withhold it. This argument will be examined as follows. Firstly, the history of the common heritage concept will be briefly examined, and secondly, the CBD will be discussed, being the 1992 convention initially providing for international governance of both human and non-human biological resources. Thirdly, the exclusion of human DNA from the CBD in 1995 is considered together with the responses of relevant international bodies, and fourthly, the profile of bioprospecting is compared between non-human and human genetic resources in order to examine the underlying argument for benefit sharing in both cases.

3.1 A History of the Concept

When should resources that occur in nature, such as air, land, the sea, water, timber, and minerals be enclosed or owned by some to the exclusion of others, and when should they be free for the communal or public benefit of all?

John Locke described the common nature of natural resources as follows:

> God [...] has given the world to men in common [...] the earth and all that is therein is given to men for the support and comfort of their being. And the fruits it naturally produces and the beasts it feeds belong to mankind in common, as they are produced by the spontaneous hand of nature: and nobody has originally a private dominion exclusive of the rest of mankind. (Locke 1690 (1988), para 26)

Under certain circumstances, including lack of abundance, common property can indeed become divided, allocated and exclusive dominion claimed. David Hume referred specifically to the sea and the air as being seemingly inexhaustible and not subject to any separate or exclusive dominion (Hume 1751). The broader term *natural resources* encompasses the range of naturally occurring resources that humankind has contested for and utilised over millennia. Military might and the ability to subjugate weaker peoples led to enhanced bargaining power, and in a process described by Jared Diamond in his analysis of the fates of human societies over the past twelve millennia, competition between peoples over resources inexorably produced the unequal distribution of affluence that is currently observed (Diamond 1997).

The 19th and 20th Centuries witnessed a continuation in the colonialist endeavours of the more powerful states, some securing entire countries as legitimate sources of natural resources such as oil, minerals, timber and other commodities (Pakenham 1991). By the mid-20th Century, when the United Nations General Assembly adopted Resolution 1803 on Permanent Sovereignty over Natural Resources confirming that states have sovereign rights over their own biological resources, the developed countries had become firmly entrenched in the advanced levels of power and wealth that define their relative status today. The common heritage framework has been described by Knoppers as one that "argues against private appropriation in favour of sharing, administration in the common interest, benefits and burdens equitably distributed, equitable access, peaceful use and preservation for future generations" (Knoppers 2003, p. 2).

The above shows that the *common heritage of humankind* is a well-known concept in international law, an attractive proposition both for those who argue against such private appropriation and for those whose priorities are the furthering of knowledge and medical science for humankind. It has a pleasing moral simplicity, and at first glance would appear to present a good argument for all human DNA to be made freely available in the interests of the common good. Each of the common heritage treaties have however commenced with similar exhortations for the subject of the treaty to be placed beyond individual appropriation, but have proceeded to develop guidelines and frameworks to ensure that the utilisation and distribution of rights are fair to all.

The *common heritage* aspiration contained in all of the above treaties was followed and superseded in each case by practical guidelines, norms and rules giving distributive effect to overarching principles of equitable sharing as set out in the Moon Treaty (UN 1979). It will be argued below that the common heritage idea which similarly was an initial aspiration for the field of genetic resources, both the non-human and human form being originally included in the CBD, was not able to endure without adjustment in the face of the competition for the resources, and in the face of the need for distributive principles to ensure equitable sharing and fairness between parties.

3.2 The Convention for Biological Diversity, State Sovereignty and Benefit Sharing

To respond further to the suggestion that human biological resources should continue to be regarded as *common heritage of humankind,* it is necessary to summarise some of the relevant controversies. Throughout the 1970s and 1980s an international controversy erupted over ownership of germ-plasm which became known as the 'seed wars' (Juma 1989), the kernel of the controversy then being described as the prevailing state of affairs when "germ-plasm flows out of the south as the *common heritage of mankind* and returns as a commodity" (Kloppenberg et al. 1988, p.10). It has been claimed that the value of plant genetic resources received over the years as free goods from the developing world has been exploitative, worth vast amounts of money to capitalist nations (Kloppenburg 2004, p. 169).

3.2.1 Bioprospecting or Biopiracy?

During the decades preceding the Rio Earth Summit in 1992 many activists attacked the term bioprospecting as being altogether too bland, masking in their view the illegitimacy of centuries of unbridled plunder. Vandana Shiva (2005) stated that the term imposed a commercial view on the developing world and, [...] assumes that prior to prospecting, the resources of desire were unknown, unused and without value. Using terminology derived from earlier "prospecting" for minerals and fossil fuels, "bioprospecting" obscures the fact that living resources are not non-renewable and are not without value prior to exploitation by global commercial interests for global markets (ibid, p. 15).

Vocal critics of bioprospecting believed that the ongoing exploitation inherent in these transactions were better described by the more derogatory term *biopiracy*, which again in the words of Shiva, refers to the use of intellectual property systems to legitimise the exclusive ownership and control over biological resources and biological products that have been used over centuries in non-industrialised countries (Shiva 2001).

Other definitions of biopiracy include criteria such as the acquiring of exclusive monopoly control through use of intellectual property and in particular patenting,[1] and the lack of prior informed consent for the transaction.[2] In this arena, indigenous peoples are amongst those most visibly aggrieved by the outcomes of this centuries-old power imbalance, having been all too often on the receiving end of the pirate-like use of their knowledge relating to plant resources. The simmering anger of this grouping is hinted at by words of indigenous activist and academic Debra Harry. She states that, "Countries that historically have been robbed of their genetic resources by more powerful states are determined to establish level playing fields rules that allow them a fair share of the benefits arising from their use of the resources" (Harry and Kanehe 2005, p. 101).

3.2.2 Bioprospecting Case Studies

The publicized allegations of biopiracy all depicted similar patterns of commodification, accompanied by patenting of promising genetic resources from the developing world. For example, a patent was taken out by *WR Grace and Co* on the well-known Neem tree of India, which was famously revoked in 2000 after community opposition on the grounds of lack of novelty and inventive step.[3] This case was widely cited as, "the world's first case against biopiracy" (Ugas 2005, p. 20) in a matter involving genetic resources of a developing country. Another example was the notorious application by the University of Mississippi for a patent in the USA for turmeric powder,[4] which was a well-known wound-healing medicine in India. This patent caused considerable anger in the country of origin, and was similarly successfully revoked after objectors persuaded the patent office that the plant use was common knowledge in India (Ganguli 2001, p. 156).

---

[1]The Action Group on Erosion, Technology and Concentration, 2005, Group web site, http://www.etcgroup.org/text/txt_key_defs.asp.

[2]Report on Biopiracy by the Edmonds Institute in cooperation with the African Centre for Biosafety; available at http://www.edmonds-institute.org/outofafrica.pdf.

[3]EPO Patent 436257 B1.

[4]US Patent No. 5401, 504 issued on 28 March 1995.

A plant related to mustard called *Maca*, cultivated and used for centuries by Andean populations for various medical purposes including sexual dysfunction, was patented by a US company *Pure World Botanicals* for sales in the US.[5] Dutfield questions whether this patent, which was not attacked or revoked, harms the Peruvian maca farmers in any way, claiming that the patent, like many others, was probably improperly awarded due to its lack of a genuinely patentable invention (Dutfield 2004, p.50). Another renowned biopiracy claim relates to the rosy periwinkle, which is native to Madagascar, amongst other countries. This plant was well known for centuries as a folk medical treatment for diabetes and other ailments. The company *Eli Lilly* researched and registered patents for treatment of leukemia in the 1950s, generating huge profits, estimated as being $100 million annually (Rosendal 2006, p. 4) by the worldwide marketing of a lucrative leukemia drug called *Oncovin* (Brown 2003, p. 136). Whilst the country of origin received no recompense, the rosy periwinkle case has been regarded by many as a classic case of exploitation (Stone 1992).

3.2.3 The International Response to Exploitation and the CBD

As the above cases show, biogenetic resources, whether from plants, animals, microorganisms or humans, were still being freely exchanged through the 1980s. Transactions were typically bilateral encounters, subject to no international supervision, with the resources being freely appropriated from the countries of origin in a manner consistent with their being the common heritage of humankind. The realization had however, dawned that bioprospecting was unjust, unsustainable and in need of regulation. Developing nations were no longer satisfied with their biogenetic resources being regarded as some form of common property, available for exploitation by the developed world. As Dutfield comments,

> […] there was a huge disparity in the way the commercial benefits from industrial use of these resources were distributed, with the lion's share going to large corporations in the developed world. Resentment about this situation made developing countries resistant to pressure from the rich countries to conserve their biological diversity at their own expense for the enrichment, as they perceived it, of these corporations. (Dutfield 2003, p. 5)

This unregulated common heritage regime was finally brought to an end by the CBD, which provided the legal framework for the exchange of biological resources worldwide.[6] The CBD was negotiated over ten annual intergovernmental meetings commencing in 1988, during which period two interrelated themes had dominated the debates. The first was that developed countries needed continued access to the natural biological resources that were chiefly situated in developing countries, and

---

[5]US Patent No. 6, 267, 995 issued on 31 July 2001.

[6]Genetic material is defined in article 2 of the CBD as "any material of plant, animal, microbial or other origin containing functional units of heredity" and genetic resources are defined as "genetic resources of actual or potential value" (CBD, Article 2).

the second was that provider states insisted that such access should no longer be deemed part of any form of global commons. In the end, it was concluded at the Rio Convention of 1992 that access to biological resources needed to be accompanied by a fair exchange in the form of benefit sharing, not only for the provider state, but also in some cases for indigenous communities.[7]

The formal objectives of the Convention were stated as being firstly the conservation of biological diversity, secondly the sustainable use of its components, and lastly the fair and equitable sharing of benefits from the use of genetic resources (CBD 1992). Benefit sharing was thus made explicit as an "evolving concept that supports the conservation of and sustainable use of the world's resources that are a collective and vital interest of all mankind" (Baslar 1998, p. 278).

3.2.4 State Sovereignty Over Resources

The CBD's unequivocal placing of sovereignty over biogenetic resources in the hands of states was regarded by many as being a turning point in international law, marking a decisive shift in the power relations of stakeholders in international trade. It should be noted however that this was not as radical as it seemed, for the United Nations General Assembly had in 1962 confirmed the general principle that states have sovereignty over their territories including the natural resources existing within them.[8] Whilst the CBD confirmed the latter fact explicitly, the words of the preamble affirming that conservation of biological diversity is a common concern of humankind nevertheless made it clear that this sovereign control was not an absolute right.

All genetic resources, which at that time were deemed to include both human and non-human biological material, were thenceforth to be managed and controlled by states in their capacity as sovereign custodians. States were thus required under the CBD framework to enact laws domestically in order to give effect to the principles and commitments contained in the convention. Not only were states obliged to respect the need for their biogenetic resources to be made available for humankind, but they were also required to respect the rights of a significant other class of stakeholders, namely the indigenous peoples within their borders.

The Nagoya Protocol on Access to Genetic Resources and the Fair and Equitable Sharing of Benefits Arising from their Utilisation to the Convention on Biological Diversity (Nagoya 2010) placed practical flesh on the previously bare bones of access, benefit sharing and compliance constituted by the CBD framework, and linked the rights of indigenous and local communities to grant access, by means of prior informed consent, to associated genetic resources. States are obliged under the Nagoya Protocol to take measures to ensure that these rights, including those of indigenous peoples, are given effect by promulgating domestic legislation.

---

[7]Article 8J of the CBD sets out the obligations of states towards indigenous Peoples.

[8]General Assembly Resolution 1803 on Permanent Sovereignty over Natural Resources.

3.3 Human DNA Governance Following Exclusion from the CBD

The human genome is an unusual and interesting form of resource; at once common to all and definitive of the human species, whilst at the same time being unique to individuals. As described above, the CBD had formulated a novel and decisive institutional response to the perceived exploitation of biological resources from the developing world. The governance and utilisation of human genetic resources in global research, being included in the definition of biological resources encompassed by the CBD, was thus entirely subject to the binding international framework described above. However, the state parties to the CBD, after a process that was not clearly captured in the public record, decided after a mere three years of the convention's existence to exclude human genetic resources from its scope. The formal resolution simply affirmed that human genetic resources are not included in the framework of the Convention (CBD 1995); thereby bringing about the regulatory and legal vacuum that informs the purpose of this book. Why then were human genetic resources excluded?

3.3.1 The First Five Years Following the CBD Exclusion

It would not be surprising to discover that the properties and distinctions that make the regulation of human DNA so different from non-human biogenetic resources contributed towards the 1995 decision. However, the following discussion makes it clear that the common heritage of humankind notion, so thoroughly rejected by the CBD, also continued in the initial years after exclusion to remain visible and explicit in the field of human genetic resources.

In 1995, the year that human genetic resources were excluded from the remit of the CBD, the Ethics Committee of the Human Genome Organisation (HUGO),[9] an international organisation of scientists involved in human genetics, issued its first statement on the *Principled Conduct of Genetics Research*. The introduction to the document referred to the primary concerns arising from the Human Genome Project, the Human Genome Diversity Project and other genetic research, singling out for emphasis the "loss of access to discoveries for research purposes, especially through patenting and commercialisation" (HUGO 1995, p. 1). The statement then laid down four principles upon which its recommendations were to be based, foremost of which was, "recognition that the human genome is part of the common heritage of humanity" (ibid, p. 1).

This first statement on governance of human genetic resources was characterised by the concern that access to the human genome should be as unrestricted as

[9]The Human Genome Organisation (HUGO) an international organisation of scientists involved in genetics, was formed in 1988 and currently represents 1 200 members from 69 countries.

possible; calling for example for collaboration between industrialised and developing countries for the reason that, "free flow, access and exchange of information is essential not only to scientific progress but also for the present or future benefit of all participants" (HUGO 1995, p. 3). It therefore seems more than likely that the reasoning underpinning this repeated aspiration, namely that open access to the human genome is essential for humankind, was behind the apparently associated decision to remove human DNA from the then untested CBD paradigm.

The exchange of human genetic resources for medical research purposes has over the past decades been to some degree governed by the non-binding guidelines of the World Medical Association's Helsinki Declaration as well as of the Council for International Organisations of Medical Sciences, (CIOMS) as will be further discussed below. These two bodies are concerned with medical research involving humans, usually in the form of research participants, but also including donors of biological samples. However, the significance of HUGO referred to above resides in the fact that it is the only international institution dedicated to human genomic research, and for that reason its further pronouncements will be examined as both reflecting and influencing the changing state of the field.

In 1997, the United Nations Educational Scientific and Cultural Organisation (UNESCO) followed HUGO's initial lead in its *Universal Declaration of the Human Genome and Human Rights* by declaring the human genome the *heritage of humanity,* emphatically endorsing HUGO's appeal in 1995 for the creation of a global commons in this increasingly important research resource. The UNESCO declaration emphasised in article 12 (a) that benefits from advances in biology, genetics and medicine, concerning the human genome "shall be made available to all, with due regard for the dignity and human rights of each individual" (UNESCO 1997). The tone and wording of this instrument also supports the notion that advances or benefits from the human genome should be common property, to be made available to all, and not susceptible for any form of private gain.

3.3.2 Benefit Sharing for Research Participants Introduced

This common heritage approach proved however, to be short-lived, failing to provide balance and a sense of justice in genomic research. For reasons closely related to the interplay between research, intellectual property rights and commodification (discussed below) a gathering sense of potential and perceived injustice in developing world genomic research influenced HUGO to urgently address the issue of benefit sharing. In 2000, the Ethics Committee of HUGO issued a comprehensive *Statement on Benefit-Sharing*, in which it stated as follows,

> The Committee believes that the issue of benefit-sharing merits further discussion because expenditures by private industry for genetic research now exceed the contributions of governments. Many new products, including vaccines and drugs for common diseases, are now based upon genetic research. Much government or non-profit research will eventually have to be commercialised. Companies involved in human health may have special moral obligations. (HUGO 2000a, part 1)

The two motivations made explicit in this statement, namely that private industry is a major driver of research on the one hand, and that new products and commercialisation are unavoidable realities on the other, comprised an acknowledgement of the dynamics outlined in the discussion on global research above. Whilst open access to a resource labelled as *a common heritage of humanity* might be a noble aspiration for global research, policy makers are required to address the stark reality which is that the entrepreneurial drive for commercial success forms an intrinsic part of the global research endeavour. Continuing in this vein, the statement describes the issue at hand as being, "whether and how to distribute profits that may accrue to *commercial enterprises, governments, or academic institutions* on the basis of the participation of particular communities" (emphasis added). The statement concludes with the words, "in the interest of justice, the last decade has witnessed an emerging international consensus that groups participating in research should, at a minimum, receive some benefit" (HUGO 2000a, p. 2).

The above extracts are significant in that they call for justice in the exchange of human genetic resources. They remind the public that particular nations or communities provide access from time to time, and refer to the commodification that motivated the benefits required in the consensus of the CBD. Finally, the HUGO statement proceeds to define and discuss key concepts such as community, common heritage, solidarity and lastly justice, all of which are central to this book. In the latter regard, the phrases *inequality between rich and poor nations*, the *direction and priorities of research*, *difference in power*, and *a possibility of substantial profit*, resonate with the debates around exploitation that preceded the adoption of the CBD. They also reflect the adoption and transference of these key-motivating concepts into the world of genetic research. Moreover, in the paragraph on benefit sharing, the HUGO Ethics Committee describes the term *benefit*, and in a clear reference to the CBD proceeds as follows, "in genetic research in general, benefit sharing has also been established as a principle of international law in the area of biodiversity and genetic resources in food and agriculture" (HUGO 2000a: 2).

After a warning against undue inducement, which is dealt with later in the book, the HUGO Ethics Committee then proposed benefit sharing in genomic research in the following terms:

1. That all humanity share in, and have access to, the benefits of genetic research
2. That benefits not be limited to those individuals who participated in such research
3. That there be prior discussion with groups or communities on the issue of benefit sharing
4. That even in the absence of profits, immediate health benefits as determined by the community needs could be provided
5. That at a minimum, all research participants should receive information about general research
6. That profit-making entities dedicate a percentage (for example 1–3 %) of their annual net profits to healthcare infrastructure and/or to humanitarian efforts. (HUGO 2000a, b, p. 2).

This forceful, albeit not binding, statement resonates with and replicates the essential elements of the access and benefit-sharing regime instituted by the CBD. On the one hand, the need for access to the resource, representing the benefit to humanity, is strongly assured in the first recommendation. However, this is balanced by the clear contention, at that stage novel in the field of human genetic resources, that benefit sharing is required at a community or provider level. Whilst only a small percentage of genomic research endeavours result in actual profits (McClellan 2003) the HUGO statement of 2000 nevertheless placed benefit sharing at the level of a requirement rather than a mere possibility, for which it received much scholarly support (Knoppers et al. 2000; Weijer 2000).

HUGO proceeded shortly thereafter to issue a further statement, namely the *Statement on Human Genomic Databases of 2002*, which again asserted common good priorities such as the global public good arising from genetic research, and the need to secure rapid access to primary genomic sequences for the benefit of humanity by placing them in the public domain (HUGO 2002).[10] This document is remarkable for the manner in which it frequently refers to the distribution of benefits amongst the wide range of stakeholders, and for its unabashed acknowledgement of the commercial realities of global research. After a range of practical recommendations for the ethical conduct of human genomic databases, the sixth detailed recommendation states that, "researchers, institutions and commercial entities have a right to a fair return for intellectual and financial contributions for databases, there should be reciprocity and exchange of information with fair return". In the same breath it reaffirms that, "any fees should not restrict the free flow of scientific information and equitable access" (HUGO 2002:4).

For those who might well have proposed that human genetic resources should continue to be governed and exchanged as part of the common heritage of humanity, the professional guidance emanating from HUGO would have given pause for thought. After the early pronouncements of both HUGO of 1995 and the UNESCO Declaration of 1997, which had declared the human genome *the heritage of humanity*, the subsequent statements and documents represented a reversal on the issue. In the absence of other explanations, it has to be assumed that it was the reality of the global research endeavour involving human DNA, namely the inherent unfairness of the bioprospecting process that led to their proposing the very same elements of a just exchange as had been advocated for in the CBD.

UNESCO subsequently issued two final documents relevant for the field of human DNA. First was the *International Declaration on Human Genetic Data*, which was formulated in 2003 by its international bioethics committee after wide consultation, aiming to guide states in the collection, processing, storage and use of human genetic data (UNESCO 2003) Article 19 of this Declaration provided a broad description of various possible benefits to groups taking part in genetic research, and specific mention is made of the need for states to collaborate, as well as to assist the capacity of developing countries to collect and process human genetic data. The need for

---

[10]Global public goods are defined in HUGO 2002 as goods, "whose scope extends worldwide, are enjoyable by all with no groups excluded, and when consumed by one individual, are not depleted for others".

benefit sharing was more evenly balanced between the requirements of humankind for benefits resulting from the use of human genetic data, and the limited medical and collective deserts of providers of samples. Secondly, UNESCO followed the above declaration on human genetic data with its *Universal Declaration on Bioethics and Human Rights* of 2005, the aims of which reflected the balance between access and reciprocity as set out in the CBD. Article 2(f) states as one of its aims,

> [...] to promote equitable access to medical, scientific and technological developments as well as the greatest possible flow and rapid sharing of knowledge concerning those developments and the sharing of benefits, with particular attention to the needs of developing countries. (UNESCO 2005, Article 2).

The need to share benefits with society as a whole, and particularly with developing countries is again emphasised in Article 15 (1), which provides a list of by now familiar forms that benefits may assume whilst falling short of the percentage of profits suggested by HUGO.[11] It should be noted that these primarily non-financial benefits are similar to those promoted by the Nagoya Protocol.

To summarise, the Ethics Committee of HUGO followed by the participating states in two subsequent declarations of UNESCO recognised that in order to secure equitable access to a valuable resource namely human genetic material in a just manner, benefits need to accrue to the provider communities as well as to humanity. The trajectory of events since 1995, as described above, presents support for the assertion that the idea of DNA as common heritage of humankind is no longer broadly supported and that considerations of benefit sharing, in one form or another, are necessary.

3.4 The Profile of Bioprospecting: Non-Human and Human Genetic Resources

We have now seen that the CBD, HUGO and states members of UNESCO all firmly proposed benefit sharing in order to bring about justice and to prevent exploitation in the research exchange, confirming after a brief hiatus that human genetic resources are no longer part of the common heritage of humankind. As far as international governance is concerned, research exchanges of both non-human and human genetic resources were by these processes decisively removed from the common heritage of humankind.

One could argue that ethical guidelines such as those issued by HUGO or the UNESCO cannot replace a valid argument in favour of benefit sharing. Hence,

[11]Article 15(1) Benefits resulting from scientific research and its applications should be shared with society as a whole and within the international community, in particular with developing countries. In giving effect to this principle, benefits may take any of the following forms, (a) special and sustainable assistance to, and acknowledgement of the persons and groups that have taken part in the research, (b) access to quality health care, (c) provision of new diagnostic and therapeutic modalities or products stemming from research, (d) support for health services, (e) access to scientific and technological knowledge, (f) capacity building facilities for research purposes, (g) other forms of benefit consistent with the principles set out in this declaration.

before proceeding on the assumption that benefit sharing is required in the exchange of human genetic resources with developing countries because HUGO says so, a further argument will be provided. The argument also has a legal base. It will be assumed that the legally binding CBD, adopted by 193 State Parties, has a significant moral force, similar to other legally binding human rights legislation. By showing the equivalence between non-human and human genetic resources in relevant respects, the force of the CBD will therefore be used to justify benefit sharing for human genetic resources. Bioprospecting is therefore under the spotlight. The two essential components of bioprospecting namely the *genetic resources* that are accessed for research, and the *utilisation* or commercialisation thereof, will be briefly compared in regard to non-human and human genetic resources.

3.4.1 The Genetic Resources, Human and Non-Human

In order to compare the two forms of genetic resources, their essential nature and properties, their value for research, and modes of access need to be examined.

The *nature and properties* of biological resources for purposes of genomic research would appear at first glance to be similar, whether performed on the tissue of plants, animals, humans, or on any of the pathogens, viruses and other microscopic organisms that are associated with the functional units of heredity of life. The fact that human DNA carries special significance associated with its personal origins, giving rise to a range of moral and spiritual issues does not affect the intrinsic structure at the microscopic level (Pullman and Latus 2003, p. 20). What is further evident from the patenting process in respect of genome sequences is the fact that at the microscopic level non-human and human genetic resources are virtually indistinguishable from one another, DNA being a universal genetic code across all species. Studies have shown how few genes are uniquely human, and as convincingly argued by Andanda et al. (2013) it is practically impossible to separate the origins of genetic resources in the innovation process. Human and non-human DNA is virtually indistinguishable in physical form.

The *value* of genetic resources for research provides a second area of comparison. An editorial for the publication *Nature*, headed genetic benefit sharing, discussed the implications of human genomic research, suggesting that the potential for injustice between developed and developing countries was equally likely, and added that, "huge profits are expected to accrue from genetic research" (Berg et al. 2000, p. 49). Commercial gain was thus accepted as the inevitable reality in human genetic research, in the same way as it had been shown in non-human genetic resources. In his book entitled *Biocapital: The Constitution of Postgenomic Life,*[12] Rajan confirms this conclusion by describing how biotechnol-

[12]Biocapital according to Rajan involves the circulation and exchange of money and commodities, but in addition a new form of currency namely biological material and information emerges. The information is decoupled from its material biological source, such as the tissue or cell line, and holds a separate value.

ogy has become a form of enterprise inextricable from contemporary capitalism (Rajan 2006, p. 3). Both forms of genomic resources aim to contribute towards drug development as a primary driver of scientific research, which process "inextricably integrates use of human and non-human resources into the same discovery programmes, even into the same molecules" (Andanda et al. 2013, p. 15).

Genetic material stored in bio-banks becomes a repository of informational wealth, described by the Norwegian Research Council as being metaphorically comparable with "gold mines" and even equated with other natural resources such as oil, waterfalls and gas (Karlsen et al. 2011, p. 572). Value is not only commercial in nature, and authors such as Barsh (1999) have described the heightened academic standing and advancement of academic institutions associated with research data in their possession as one of the broader forms of value that flow from successful genetic diversity research with indigenous populations (ibid, p. 375). As a final component of value, Andanda describes a range of examples of what the authors term the *Inextricable Utilisation of Plants, Human Genetic Resources and Microbes in Drug Development* (Andanda et al. 2013, p. 15). For example, a drug target for malaria was discovered in one study using human samples, whilst another study arrived at similar conclusions using samples from Shitake mushrooms native to East Asia (ibid, p. 17). Genetic data of both human and non-human origin are lucrative even before medical products have been developed (Mackintosch 2005, p. 217) but it is the existence of patents, discussed below, that boosts commercial value.

Access to genetic resources is a final realm of comparison, human and non-human resources being similarly and easily obtained. No more than one tiny sample is required, and once the genetic information is encoded, access is complete and the original sample is no longer required. Drawing together this analysis of genetic resources in research, described as the first essential component of bioprospecting, it is clear that human and non-human genetic resources are virtually indistinguishable in nature, in their essential value for research, and also in regard to their general means of access from indigenous or vulnerable communities in the developing world.

3.4.2 The Utilisation of Genetic Resources

The second component of bioprospecting encompasses *utilisation.* A large proportion of research and development for genetic research is funded by the private sector, which seeks commercial rewards in addition to useful discoveries that serve public health. In order to understand utilisation, we will need to examine briefly the intellectual property rights regime and biotechnology patents, the latter described by John Meyer as representing, "the most dramatic step to date in the commoditisation of life itself" (Meyer 2000, p. 156). In this process, we examine whether in this second component of bioprospecting, there is any logic behind the differentiation of non-human and human genetic resources. This brief analysis will commence with an introduction to the intellectual property system, which operates for the protection of innovation generally, including the role of patents and persistent attempts to ensure open access to the results of scientific research.

(a) The intellectual property system

The underlying purpose of all bioprospecting is the scientific quest for and the acquiring of knowledge, which in some cases results in what are termed "intellectual property rights" to the important genetic resource fruits thereby harvested. It is these intellectual property rights (IPRs) to knowledge, defined simply as *creations of the mind* (Wunsch Vincent 2012), that enable ownership and commercial exploitation of research discoveries, and which require further scrutiny in the context of genomic research.

IPRs can be owned and protected like any other form of property, and are designed to protect human creativity and inventiveness by means of a range of specific mechanisms which include patents for all forms of inventions, copyright for the form of a creation, as well as trademarks, industrial designs and others. IPRs are in the words of the World Intellectual Property Organisation[13] (WIPO) at the heart of business strategies, and IPR policy "has moved to the forefront of innovation policy" (ibid, p. 1). Demand for patents has grown from a global estimate of 800,000 applications in the early 1980s, to an estimated 2 million in 2010, whilst global trademark and industrial design applications over the same period have risen from 1 million and 290,000 per year to 3.7 million and 650,000 respectively (ibid, p. 12).

Patents, as the primary instruments for the protection of human inventions, grant monopoly rights for a fixed number of years to the inventor[14] in the country of registration. Applicants must satisfy a national patent-issuing authority that the invention, process, device, idea, compound or subject meets three criteria. These are *novelty*, *non-obviousness*, that is, it must contain an inventive step and not be an obvious variation of a known technology and *usefulness* in that it is capable of being applied (ibid, p. 2).

The peculiar aspect of the right is that it relates to information, which can be incorporated in tangible objects, and therefore results in control over products and markets (South Centre 2000, p. 1). The IPR regime had its origin in the Paris Convention for the Protection of Industrial Property of 1883 and it developed predominantly in the Western economies of the USA, Europe and Japan. It was universalised by means of the World Trade Organisation (WTO), whose Trade Related Aspects of Intellectual Property Rights (TRIPS) agreement of 1994 had as its purpose to provide a uniform international framework for the promulgation of IPRs within individual member states (Maister et al. 2012, p. 17).

Many authors have challenged the fairness of TRIPS (Correa 1998) with some opposing the impact in particular on indigenous peoples (Tauli-Corpuz 2003; Harry and Kanehe 2005) and also opposing the patenting of life (Shiva 2001, p. 41). Others object primarily to the dominance of the economies of the

[13]The World Intellectual Property Association, based in Geneva, was formed in 1974 as a specialized agency of the United Nations, with the primary task of coordinating intellectual property issues worldwide.

[14]This is usually 20 years.

industrialized North (Correa 2005) and others still object to the impact on prices of patented essential drugs in developed countries (Maister et al. 2012, p. 28). Schroeder and Singer (2010), examine the role of the IPR system in hampering access to and availability of life saving drugs in the developing world, concluding that the death of 10 million people each year due to high monopoly prices and lack of access to life-saving drugs is one of the "stains on humanity's conscience" (ibid, p. 9).

For patents based on genetic discoveries, it makes no difference whether the research process involved human or non-human DNA. And the exploitation claims of developing countries as well as the criticism of the patent system are also equivalent in respect to both forms of DNA.

(b) Open access
One could ask whether the patenting system is not losing ground, given the increasing emphasis on open access. Open access has long been an aspiration in opposition to the commodification process described above. The resolve to protect the public domain from the privatisation and commercialisation associated with patents has remained a powerful theme in publicly funded scientific research, such as the *Human Genome Project*, the *SNP consortium*[15] and the International HapMap project (Rajan 2006, p. 15). The *Bermuda Rules*[16] (Cooke-Deegan 2010, p. 19) were intended to ensure rapid release of sequencing data into the public domain, and to thereby prevent sequence-based patents. Patent pools, clearinghouses, exchanges, open source models and liability regimes are all part of the drive to harness or mitigate the unbridled workings of the profit motive (Wunsch Vincent 2012). Efforts to reverse privatisation are thus numerous, under the influence of motives described as defending the public domain (Lange 1981), or creating a new commons (Benkler 2004).

A WIPO report on the changing face of innovation, however, concludes that despite evidence of a greater openness and collaboration in the innovation process, IPR ownership nevertheless remains at the forefront of business strategies (Wunsch Vincent 2012). This confirms a *proprietary* business model,[17] which is founded on the normative belief that science and innovation are best served by IPRs and financial profit (Ziman 2002). Many authors agree with Ziman that commercialisation is here to stay (Reddy 2005, p. 449; Raj and Eisenberg 2003, p. 310), and that despite the open access movement, privatisation of knowledge remains a core component of scientific research.

---

[15]The SNP consortium, a public-private partnership consisting of universities and private companies, devised an elaborate IPR process to avoid a scramble for patent rights relating to single nucleotide polymorphisms (SNPS), and to ensure that the discovery and characterization of SNPs remained in the public domain.

[16]The Wellcome Trust spearheaded a 1996 meeting in Bermuda between the high-throughput sequencing centers, who agreed inter alia to make sequence data available within a day, once a contiguous stretch of 1000 nucleotides had been assembled.

[17]According to Ziman the proprietary model is guided by the principles Proprietary, Local, Authoritarian, Commissioned, and Expert science.

To conclude this comparison of the profile of bioprospecting, it is apparent not only that genetic *resources* of human and non-human origin are entirely alike in nature, form and inherent value, but also that their *utilisation* in research follows an entirely similar trajectory. In the latter regard, the intellectual property system is an essential component in the privatisation of research discoveries, in a commercial reality that persists despite all efforts to bring about open access. This examination of the profile of bioprospecting has shown the close similarity between human and non-human DNA both in their nature and their utilisation, enabling the conclusion that benefit sharing is appropriate in order to prevent exploitation in the research transaction.

3.5 Conclusion

This chapter had as its overall purpose an examination of the argument that human genetic resources should continue to be governed as part of the *common heritage of humankind*, and that human DNA should consequently be free from considerations of benefit sharing.

Perceptions of exploitation and unfairness in the exchange of non-human genetic resources had led the nations of the world to remove these resources decisively from the previously unregulated common heritage regime, and to institute fair and equitable benefit-sharing as a mechanism to correct the imbalance, notably in respect of both human and non-human genetic resources. Benefit-sharing, being one of the novel and important stipulations contained in this binding international convention, was thus introduced by the CBD as part of international law in 1992.

Benefit sharing for human DNA was after a hiatus of some years first proposed as a mechanism by HUGO in 2000, followed by UNESCO and other international institutions, albeit in the form of research ethics guidelines and not in a legally binding international convention. These guidelines unequivocally reflected the inherent value of human genetic resources, and proposed consideration of various forms of benefit-sharing in exchange for their access, consistent with their decisive removal as a common heritage resource.

In order to deal more thoroughly with the original objection to benefit sharing described at the commencement of this chapter, it was required to go one step further: namely to analyse whether the profile of bioprospecting that had motivated the institutional responses set out above, was comparable to that pertaining to the human form. If so, the legally binding provisions of the CBD should apply to human DNA.

Regarding the resource, it was found that the nature of both forms of genetic resources is inextricably comparable and virtually indistinguishable at the microscopic level. In addition, the value and importance of these resources for scientific advancement were shown to be similarly significant and to a large degree interchangeable, whilst as a final point of congruence both forms of genetic resources are similarly accessed, in many cases from indigenous or rural communities situated in the developing world. Similarly utilisation of both forms of genetic resources within the parameters set by the intellectual property system showed that privatisation of knowledge for commercial gain remains the central component of scientific research for both types of genetic resources, despite the impressive advances of the open access movement to create a scientific commons. Calls for human DNA to continue to be regarded as a *common heritage* resource, therefore, disregard the realities of the profit motive and IPRs in the world of genomic research, and have in the view of this book been correctly ignored by the prevailing research ethics guidelines. It is suggested on the basis of the arguments presented above that the *common heritage of humankind* notion no longer applies in any meaningful way to human genetic resources.

References

Andanda, P., Schroeder, D., Chaturvedi, S., Meghesa, E. & Hodges, T. (2013) Legal frameworks for benefit sharing: from biodiversity to human genomics. In D. Schroeder & J. Cook (Eds.) (2013) *Benefit sharing: From biodiversity to human genetics*. (Chapter 3 p. 33) London: Springer.

Barsh, R. L. (1999). Indigenous Knowledge and biodiversity in Indigenous peoples, their environments and territories. In D.A. Posey (Ed.) *Cultural and spiritual values of biodiversity* (pp. 73–76) London and Nairobi: CIT Publications and UNEP.

Baslar, K. (1998). *The concept of the common heritage of mankind*. Boston: Martinum Nijhoff Publishers.

Benklar, Y. (2004). Commons-based strategies and the problems of patents. *Science, 305*, 1110–1111.

Berg, J. W., Applebaum, P. S., Lidz, C. W., & Parker, L. S. (2000) (eds). *Informed consent: legal theory and clinical practice*. 2nd edn. New York, NY, USA: Oxford University Press.

Brown, M. F. (2003). *Who owns native culture?*. Massachusetts: Harvard University Press.

CBD. (1992). *Convention on biological diversity*. http://www.cbd.int/convention/.

CBD. (1995). Convention on Biological Diversity Decision II/11. *Access to Genetic Resources*, Montreal: UNEP/CBD

Cook-Deegan, R., & Heaney, C. (2010). Patents in genomics and human genetics. *Annual Review of Genetics and Human Genetics., 11*, 383–425.

Correa, C. (1998). *Implementing the TRIPS agreement: General context and implications for developing countries*. Penang, Malaysia: Third World Network, Online Bookstore. http://www.twnside.org.sg/title/trip2-cn.htm.

Correa, C. M. (2005). Pro-Competetive measures under the trips agreement to promote technology diffusioon in developing countries. *Journal of World Intellectual Property 10/2005 Patenting Human DNA: What Flexibilities Does the TRIPS Agreement Allow?, 4*(4):481–496. doi: 10.1111/j.1747-1796.2001.tb00123.x.

Diamond, J. (1997). *Guns germs and steel: the fates of human societies*. New York: WW Norton and Company.

Dutfield, G. (2003). *Intellectual Property Rights and the Life Science Industries: A Twentieth Century History*. Ashgate.

Dutfield, G. (2004). *Intellectual property*. Biogenetic Resources and Traditional Knowledge: London, Earthscan Publications Limited.

Ganguli, P. (2001). *Intellectual property rights*. New Delhi Tata McGraw Hill: Unleashing the knowledge economy within.

Harry, D., & Kanehe, L. (2005). The BS in access and benefit sharing (ABS) critical questions for indigenous peoples In B. Burrows (Ed.), *The catch. Perspectives in benefit sharing* (pp. 88–120). Washington: Edmonds Institute.

HUGO. (1995). HUGO Ethics Committee. *Statement on the Principled Conduct of Genetic Research.* http://www.hugointernational.org/Statementonprincipled conduct of human genomic research.

HUGO. (2000a). HUGO Ethics Committee. *Statement on benefit sharing*. Retrieved April 9, 2000 from http://www.hugointernational.org/Statement_on_Benefit_Sharing.htm.

HUGO. (2000b). Social, Ethical and Legal Issues Committee. *Statement on the principled conduct of genetic research.* London: HUGO. http://www.gene.ucl.ac.uk/hugo/conduct.htm.

HUGO. (2002). *Human Genome Organisation statement on Human Genomic Databases.* Retrieved December 2002 from http://www.eubios.info/HUGOHGD.htm.

Hume, D. (1751). *Enquiries concerning the principles of morals.* In L.A. Selby Bigge (1978 edn). Oxford: Oxford University Press.

Juma, C. (1989). *The gene hunters: Biotechnology and the scramble for seeds.* Princeton University Press: Princeton.

Karlsen, J. R., Solbakk, J. H., & Holm, S. (2011). Ethical endgames: Broad consent for narrow interests; open consent for closed minds. *Cambridge Quarterly of Healthcare Ethics*, 20 (4), 572–583.

Kloppenberg, J., & Kleinman, D. (1988). Seeds of controversy: National property versus common heritage. In J. Kloppenberg (Ed.), *Seeds and sovereignty: The use and control of plant genetic resources* (pp. 1–15). London: Duke University Press.

Kloppenberg, J. (2004). *First the seed: the political economy of plant biotechnology 1492–2000.* Madison: University of Wisconsin Press.

Knoppers, B. M., Chadwick, R., Takebe, H., et al. (2000). HUGO urges genetic benefit sharing. *Community Genetics, 3*(2), 88–92.

Knoppers, B. M. (2003). Perspectives: International law and ethics, HUGO experience. *Toward and understanding of benefit sharing*. March 3, Philadelphia, PA.

Lange, D. (1981). Recognizing the public domain. *Law and Contemporary Problems, 44*, 147–173.

Locke, J. (1690) (1988). *Two treatises on government.* In P. Laslett (Ed.), Cambridge: Cambridge University Press.

MacIntosh, C. (2005). Indigenous self-determination and research on human genetic material. *Health Law Journal 2005.* 13, 213–250.

Maister, B., van Woensel, C., & de Jonge, B. (2012). *Harnessing intellectual property rights for development countries: The double role of IRP's in the context of facilitating MDG's No 1 and 6.* Wolf Legal Publishers. www.wolfpublishers.com/harnessingipr/pdf.

McLellan, MB. (2003). *Lecture BIO 2003.* Washington, DC. http://www.bio.org/events/2003/meida/Mclellan_0623asp.

Meyer, J. (2000). The right to life? on nature, property and biotechnology. *The Journal of Political Philosophy, 8*(2), 154–175.

Nagoya Protocol. (2010). *The Nagoya protocol on access to genetic resources and the fair and equitable sharing of benefits arising from their utilization (ABS) to the convention on biological diversity.* Retrieved October 29, 2010 from http://www.cbd.int/abs/about/.

Pakenham, T. (1991). *The scramble for Africa: white man's conquest of the Dark Continent from 1876 to 1912.* London: Harper Collins.

Pullman, D., & Latus, A. (2003). Benefit sharing in smaller markets: the case of newfoundland and labrador. *Community Genetics, 6*, 178–181.

Raj, K., & Eisenberg, R. S. (2003). Bayh-Dole reform and the progress of biomedicine. *Law and Contemporary Problems*, 66(1–2) 289–314. http://www.law.duke.edu/shell/cite.pl?/66+Law+and+Contemp+Probs.+289+(WinterSpring+2003).

Rajan, K. (2006). *Biocapital: the constitution of postgenomic life.* Durham: Duke University Press.

Reddy, D. S. (2005). Good gifts for the common good: blood and bioethics in the market of genetic research. *Cultural Anthropology, 22*(3), 429–432.

Rosendal, G. K. (2006). Regulating the use of genetic resources-between international authorities. *European Environment, 16*(5), 265–277.
Schroeder, D., & Singer, P. (2010). Access to life-saving medicines and intellectual property rights: an ethical assessment. *Cambridge Quarterly of Healthcare Ethics, 20*, 1–11.
Shiva, V. (2001). *Protect or plunder?* Understanding Intellectual Property Rights. London: Zed books.
Shiva, V. (2005). Bioprospecting as sophisticated biopiracy. In B. Burrows (Ed.), *The catch; perspectives in benefit sharing* (pp. 15–37). Washington: The Edmonds Institute.
South Centre. (2000). *The TRIPS agreement: A guide for the South.* Geneva, Switzerland: South Centre. http://apps.who.int/medicinedocs/en/d/Jh3009ae/1.html.
Stone, R. (1992). The biodiversity treaty: pandora's box or fair treaty? *Science, 256*, 1624.
Tauli-Corpuz, V. (2003). *Biodiversity, traditional knowledge, and rights of Indigenous peoples* (p 33). Penang, Malaysia: Third World Network. http://www.twnside.org.sg/title2/IPR/pdf/ipr05.pdf.
Ugas, R. (2005). *Policy development in the organic movement.* Leisa Magazine, 21(4). http://ileia.fourdigits.nl/magazines/global/practice-and-policy/policy-development-in-the-organic-movement.
UN. (1979). *Agreement governing the activities of states on the moon and other celestial bodies.* New York, NY: United Nations General Assembly. Resolution 34/68. http://www.oosa.unvienna.org/oosa/SpaceLaw/moon.html.
UNESCO. (1997). *Universal declaration on the human genome and human rights.* United Nations Educational Scientific and Cultural Organization. http://www.undocuments.net/udhghr.htm.
UNESCO. (2003). *International declaration on human genetic data. Human Rights.* United Nations Educational Scientific and Cultural Organization 32nd Sess, 32 C/29 Add 2. Retrieved October 17, 2003 from http://www.unesco.org/confgen/2003/genetic.
UNESCO. (2005). *Universal declaration on bioethics and human rights (1997). Human Rights.* United Nations Educational Scientific and Cultural Organization Article 2(f). http://www.unesco.org/new/en/social-and-human-sciences/themes/bioethics/human-genome-and-human-rights/.
Wang, Y. (2011). *Group protection in human population genetic research in developing countries: The People's Republic of China as an example.* Unpublished PhD book. University of Glasgow. http://theses.gla.ac.uk/3005/1/2011WangYuephd.pdf.
Weijer, C. (2000). The ethical analysis of risk. *Journal of Law, Medicine and Ethics, 28*, 344–361.
Wunsch Vincent, S. (2012). *The Changing Face of Innovation.* WIPO publication Retrieved February 2012, from http://wipo.int/wipo_magazine/en/2012/01/article_0006.html.
Ziman, J. (2002). *Real Science.* Cambridge: Cambridge University Press.

Chapter 4
The Altruism Argument

Abstract The second argument against benefit sharing for human genetic resources holds that humans should freely offer their DNA for research out of a sense of altruism, without expectation of reward. Proponents of this argument suggest a duty to participate in research, rather than a right to benefit in any way. The chapter examines this argument by tracing the history and meaning of the term altruism through the literature, distinguishing for example between *beneficence,* namely doing good, and *benevolence,* which is wishing good. The notion of *reciprocity,* as one that is said to encourage if not motivate the altruist, is examined from the seminal work of Marcel Mauss to the more recent work of Titmuss. *Solidarity* is then discussed as a strongly associated notion, which is widely used yet has numerous and different meanings. Finally the altruism argument is examined in the light of the *fair exchange* model of Schroeder and Gefenas, which holds that such a call is altogether misplaced in the developing world. The chapter concludes that some form of reciprocity is required to avoid Mayer's second form of exploitation, which could well be in the form of benefit sharing.

Keywords Altruism · Reciprocity · Beneficence · Benevolence · Solidarity · Fair exchange model

The argument which can be referred to as the *altruism argument* objects to any benefit sharing with human research participants for the broadly stated reason that it is simply inappropriate if not morally wrong for humans to benefit from their DNA. This argument would deny that there is a need to address the potential of exploitation of research subjects, and would claim on the contrary that humans should be altruistic in providing samples of their tissue for the common good: they owe such a duty to the public as part of their solidarity with fellow human beings.

Proponents of this argument have challenged the notion of a right to benefit sharing for research participants, suggesting something closer to a duty to participate in genetic research. In an article entitled, *Solidarity and Equity: new ethical frameworks for genetic databases*, the authors argue in favour of human participation in research for the benefit of others, stating as follows,

R.S. Chennells, *Equitable Access to Human Biological Resources in Developing Countries*, DOI 10.1007/978-3-319-19725-8_4

> [...] it is not obvious, however, why a right to refuse to participate in genetic research, when it could be to the benefit of others, should be overriding. On the contrary, it could be argued that one has a duty to facilitate research progress and to provide knowledge that could be crucial to the health of others. The principle of solidarity would strongly contradict a view that no research should be conducted if it would not directly benefit those participating in a study. (Chadwick and Berg 2001, p. 320)

Karen Berg supports this notion in an article dealing with the ethics of benefit sharing. After first stating that a nation, company or person should not make money on another's resources without paying for them, he goes on to question the rights of individuals to shared benefits, holding that they should participate on the basis of an ethical principle of solidarity with their own group, which should be a "generous gift to science" (Berg 2001, p. 242).

The above sentiments contain an attractively simple postulate in opposition to benefit sharing. The altruism argument states that human beings, out of a sense of altruism and solidarity with each other, should contribute to research for the common good, without any expectation of return. In order to address this argument I shall examine the related concepts firstly of altruism, and then secondly of solidarity. Finally, the altruism/solidarity argument will be discussed in order to assess its applicability to genomic research.

4.1 Altruism

Altruism, in its broadest sense meaning, but not limited to, promoting the interests of the other (Scott and Seglow 2007, p. 1), is an interesting concept. At its essence it is a simple, or elemental moral idea, regarding the interests of the other as one's own, and according to Scott and Seglow is often identified with the golden rule framed by Hobbes as, "do not that to another, which thou thinketh unreasonable to be done by another to thy selfe" (ibid, p. 2). Presumably one's wish for others to be altruistic towards one would, in this sense, motivate one's own altruism towards others. Under scrutiny, whether such actions would be deemed to qualify as altruism, as well as many other tricky questions of morals and ethics emerge. For example, would it be morally questionable, in addition to being altruistic, to agree to donate one's blood only to those of your own race?

Scott and Seglow explore many aspects of such deceptively simple questions, including the place of the competing drives of self-interest and egoism in humans (ibid, p. 14). The economist Adam Smith stated that egoism, or self-interest rather than altruism was what would lead to the general welfare, stating that it was not from, "the benevolence of the butcher, the brewer or the baker that we expect our dinner, but from their regard to their own interest" (Smith 1776, 1976 edition pp. 26–27). Auguste Compte, on the other hand, who invented the term "altruism," believed that altruism was centrally about promoting other people's interests, morality being the triumph of altruism over egoism (Scott and Seglow 2007, p. 15).

Kant also provides useful guidance on the morality or motive behind altruism. Kant distinguished between beneficence, which is understood as doing good, and benevolence, which is wishing good, beneficence being benevolence in action and acting in accordance with a, "maxim of making other's happiness one's end" (Kant 1797, 1996 edition Chap. 6). Whilst this appears to be noble in essence, according to Herman, Kant regarded it as a duty to be beneficent, for the reason that it would not be rational to not do so, and thereby risk not receiving similar assistance from others in the future. According to Herman, Kant regarded benevolence as an imperfect duty, namely a duty to act only by maxims that we would *wish* to be universalised. Benevolence according thus carries a lesser obligation than a perfect duty, and this scenario of individuals mutually acknowledging their human needs and duties can be termed a duty of mutual aid (Herman 1993).

This brings the discussion to reciprocity, which according to the latter argument is what to some degree at least encourages the altruist, even though it might provide a less than noble motive for her good deeds. Reciprocal altruism is altruism that is performed with the hope of obtaining a future reward from the person one benefits, in keeping with the above interpretation of the Golden Rule, and is something of a hybrid between altruism and self-interest. You scratch my back and I will scratch yours, is a term for this somewhat conditional offering of assistance, which many would claim falls short of altruism. Scott and Seaglow describe a range of other forms or definitions of altruism which include reciprocity as well as forms of other-regarding behaviour without the need for self-sacrifice, all of which forms are potentially compatible with the notion of benefit sharing proposed in this book. (Scott and Seglow 2007, pp. 21–38).

Reciprocity was famously examined by Marcel Mauss in his anthropology classic *The Gift* (Mauss 1950, 2002 edition) in which he examined gift-giving from ancient times in indigenous tribes, as well as in the more recent legal systems such as those of the Roman, Semitic, Germanic and other Indo-European societies. The seemingly ubiquitous practice of gift giving in all these societies existed separately from commercial transactions. Mauss defined a gift as, "a voluntary, unrequited, surrender of resources" (Mauss 1950, p. 3). The apparent generosity of the gifting practices seemed to indicate levels of solidarity, charity and trust far greater than identifiable today, but he famously concluded on the contrary that in all such societies there are no free gifts.

The giving of gifts engages the giver and the receiver alike in finely woven if implicit obligations and commitments that reflect and resonate with the institutions of the day. Morality did not enter the transaction, and the society's (unwritten) norms and expectations framed what was required in certain circumstances. Mauss established that the entire notion of a free gift is based upon a misunderstanding of the nature of such a transaction, and concluded that a gift that expects no return, that does nothing to enhance solidarity, is a contradiction in terms (Mauss 1950, p. xii). Mauss's work encourages us to consider that material things, whether sold or given, always retain something of the identity of the giver, and always require reciprocation of some form. This notion surely applies to a tissue sample, which retains its original identity, in the sense that it can be traced back to the original donor, long after it has been reduced to scientific forms of digital and other non-physical information.

Finally, the work of Richard Titmuss added significantly to the understanding of altruism. In his book, *The Gift Relationship: From Human Blood to Social Policy*, he attempted to counter policies that promoted the commoditization of blood, in which donors would be remunerated with a fee. Invoking anthropologists such as Durkheim, Mauss and Levi-Strauss, the central question that he sought the answer for in the process of developing his theory was, why do people give to strangers? His primary aim was to motivate for voluntary blood donation, in which ordinary people had the moral choice to give blood as a "symbolic gift of life to an unnamed stranger" (Titmuss 1973, 1997 edition p. 140). What was particularly altruistic was that the gift of blood was not to known individuals in a close relationship, as in Mauss's societies, and that the donor's reward, apart from a cup of tea and a biscuit, was purely in knowing that she had contributed towards the public good.

Reciprocity, however, seems to be present in all communities, in some or other form. The unnamed strangers that would 1 day benefit were likely to be citizens, perhaps even relations, and the donor's contribution was towards the public health of her own country. According to Scott and Seglow, who support the latter notion of reciprocity, rather than regarding the donation of blood as a true gift that expects no return, such donation should be understood as an act of creative altruism, namely an essentially reciprocal exchange relationship that fosters a sense of community (Scott and Seglow 2007, p. 111). In other words, the fact that Titmuss identified the benefit of choosing to give to unnamed strangers as being a valuable right in itself, shows that this right had the capacity to provide the donor with some of the satisfaction, which in turn is a form of reciprocation, that motivates otherwise selfless acts of altruism.

Seeking empirical support for the motivation behind blood donations, a recent study on 12 bio-banks in six developed countries indicated that donors were primarily motivated by altruism, trust and a sense of duty (Johnsson et al. 2010). As is discussed below, the form of altruism espoused by Titmuss, whereby people are encouraged to donate blood for little overt reward other than the knowledge that one is choosing to give for the benefit of humankind, is thus still relevant and applicable to research participants in the affluent or developed world. To what extent it can be evoked for developing world or less affluent research participants however remains an open question, which will be addressed later.

4.2 Solidarity

The association of the term *solidarity* with altruism is at once understandable, yet potentially misleading. In the words of Pensky (2009), solidarity is a strong candidate for the most challenging of all concepts in modern political thought, functioning across a startling array of discourses. It is invoked as an essential component of community, as the political value without which freedom becomes hollow, and is strongly associated with the 18th Century republican term of fraternity, intended as a sibling to the ideals of liberty and equality (ibid, p. 1). In the fields of moral philosophy and normative ethics, solidarity can refer to the concept of membership in a moral or other community, or the collective bonds that hold autonomous moral agents together. The origin of the term is traced to Roman law, in which the term *obligatio in solidum* defined the assumption of joint liability for a financial debt, an obligation often taken to denote a person's moral virtue in being prepared to act in support of the extended family (ibid, p. 6).

Seeking a more practical application of the term, Paul de Beer describes solidarity as the willingness to help others or to support the group one belongs to, without immediately getting something in return, and then defines it more generally as, "the positive bond between the fates of different people" (De Beer and Foster 2008, p. 15). There are many interpretations and meanings of solidarity, he warns, for example the term *social solidarity* is distinguishable from related concepts such as *social cohesion* and *social capital,* but it is *solidaristic acts* that in his view are more important than mere *solidaristic attitudes* (ibid, p. 18). He goes on to state that,

> The defining characteristic of a solidaristic act is that there is no equivalence between what one contributes to others or to the group as a whole and what one gets in return. Those who are best off will generally contribute the most, those who are worst off will benefit from the others. Acts of solidarity thus reduce the gap between the fortunate and the unfortunate (ibid, p. 18).

Solidarity could, for example, have been simultaneously used both as an argument why the Tristan da Cunha islanders should, or why they should not have received any benefits for their valuable donations of DNA samples. Solidarity with the islanders by the rest of the world would have motivated for a system that ensured their inclusion in the benefits, whereas the islanders' solidarity with humankind would have suggested that they should have expected no recompense in respect of the benefits bestowed upon the world. Using the term solidarity broadly without defining which particular act or form is meant can thus be misleading and result in entirely different conclusions.

Solidarity has somewhat predictably been used as an argument both for and against benefit sharing. As stated in an editorial for *Science* entitled, *Genetic Benefit Sharing*, solidarity was originally used as an argument for, rather than against benefit sharing. According to Knoppers, "there are … fundamental arguments in favour of benefit sharing. In the interests of human solidarity, we owe each other a share in common goods, such as health" (Knoppers 2000, p. 49). The reason provided in this editorial calls for recognition of the vulnerability of the less powerful part of humanity, implying an altogether different component of solidarity, and requiring benefit sharing as a form of corrective action in order to bring about justice,

> When there is a vast difference in power between the organisation carrying out research and the people providing material for that research, and when the organisation stands to

> make a substantial profit (albeit taking the risk of investment) concerns about exploitation arise that benefit sharing can address. Considerations of justice require action to meet basic health care needs. (ibid, p. 50)

This begs the question whether solidarity, in the form that implies the need for altruism by humankind in the arena of research, provides any sound reason why research donors from the developing world should not anticipate any form of reciprocation, or benefits?

4.3 The Altruism/Solidarity Argument Examined

Calling for solidarity under circumstances where it is in fact exploitative would be morally unacceptable, in the same way that one should not demand food in the name of solidarity from a starving person. The authors quoted at the outset of this chapter (Chadwick and Berg 2001), who demanded solidarity in genetic research, would therefore need to show that the exploitation warned against in the *Science* editorial is not a concern, and that, to apply the test for Mayer's second class of exploitation, the providers of such DNA would in all the circumstances be benefited sufficiently. No call for solidarity or altruism could succeed in the modern world in the face of knowledge that those being called upon to donate are being knowingly exploited.

At the same time, one might well question why both forms of sharing should not co-exist, and why sharing of benefits with humankind out of altruism and solidarity with all should exclude consideration of benefits to participants. It is beyond dispute that the general call for altruism and solidarity is a worthy plea, resonating with the idealistic motivations for creative altruism so persuasively expressed by Titmuss, and one that in a perfect world, or even in the wealthy countries, would be difficult to refute. It is proposed to examine and respond to this issue by answering the question, would such a call for altruism and solidarity suffice to satisfy the concerns of research participants in the developing world, or would such exchanges of human genetic material without compensation be exploitative?

4.3.1 Is a Call for Altruism and Solidarity in Research Appropriate in the Developing World or Is It Exploitative?

It seems clear that the call for altruism and solidarity resonates with the common heritage of humankind notion, as well as with associated calls for free exchange of and open access to research data in the interests of global health. However, both of these visions are in tension with opposing arguments referred to above which point out the inherent differentials in health and wealth between the parties to the research exchange, and which propose benefit sharing in order to bring about justice. The perspective being developed in this book, which takes note of the potential for exploitation of developing world research participants in the context of a highly unequal world order, is that the settings and overall conditions for potential research participants in the developing and developed world are far from comparable. Attempts to alleviate this disparity in basic health and other indicators need to take into account the idea of a non-ideal moral theory as espoused by Wertheimer.

Simm addresses this point by acknowledging the realization in the developing world that contributions made by these countries to research, namely by the provision of sought after genetic resources, are not matched by any identifiable form of altruism from the research/medical industry side; for that reason she describes what she terms an emerging consensus that some form of compensation is required to bring about a sense of justice (Simm 2005, p. 8). This point resonates with the notion of *wrongful gain*, which Mayer described as exploitation at its most basic level: this occurs when acquisition or advantage is received by one party, whilst the other party is used instrumentally and in some manner wrongfully (Mayer 2007, p. 139).

Gaining at another's expense is at the heart of this type of exploitation, which requires the loss by the one party to be in some moral manner undeserved (ibid, p. 140). As described above, it was the widely acknowledged failure in reciprocity during exchanges of genetic resources that had motivated states to formulate the benefit-sharing mechanisms of the CBD, in order to bring about a more fair and equitable situation. The kind of justice referred to by Simm is termed justice in exchange, which requires reciprocity in an exchange between two parties, rather than what is termed distributive justice, which takes a broader view of and assesses the fairness of distributions of costs and benefits generally (Schroeder and Pogge 2009).

Schroeder and Gefenas examine what they term the existing *fair exchange model* that pertains between the health care industry and human research participants in the affluent or developed countries. Most citizens in these countries benefit from a health care industry which produces new health products and services tailored to their health needs, increased knowledge about health which is conveyed to all, increased jobs in health and related sectors, and profits for research and pharmaceutical companies (Schroeder and Gefenas 2010, p. 4). People in these countries experience an inherent and tangible form of reciprocity for their participation in the complex social and economic network encompassed by the health care system, reminiscent of a Maussian society, ensuring the fairness of the entire exchange. An analysis of the Swedish general public for example, found that 86 per cent of those interviewed would freely donate blood samples for bio-bank research, motivated by altruism and their support for and trust in the public health system (Kettis et al. 2006).

Donations of human tissue for research purposes in developed countries, often described as being altruistic in nature, take place within an entirely different paradigm to that pertaining to the developing world, where the tissue donor seldom receives any form of reciprocation such as nationally funded health care systems.

It is therefore difficult to imagine how calls for altruism can be appropriate for donors of human tissue from developing world communities. This discussion has shown that no forms of exchange stand alone or free from the need for Maussian forms of reciprocity, which leads to the more specific question whether such exchanges motivated by calls on altruism are exploitative.

More succinctly put, would research donors from the developing world be sufficiently benefited, within the meaning of Mayer's second class of exploitation, if they were persuaded to donate their tissue freely to developed world data bases out of a call for solidarity? Consider the case of the Tristan da Cunha islanders, referred to above, who entered into an agreement to provide their DNA for research into the genetic causes of asthma, in exchange for free access to any possible future cure. The islanders' donations of their DNA samples to the company were solidaristic in the most general sense, in that the research was ultimately dedicated towards a possible cure that might have benefited humankind as well as the islanders. The islanders thus acted in accordance with the proposition contained in Chadwick and Berg's paper that they should donate their samples out of altruism and solidarity with others, and out of their "duty to facilitate research progress and to provide knowledge that could be crucial to the health of others" (Chadwick and Berg 2001, p. 320). However, despite such altruism on their part, the company sold the gene-based discovery for $70 million without any compensation to the original donors (Cunningham 1998, p. 18). The lack of reciprocation clearly indicated exploitation of Mayer's second class.

Somewhat less dramatic examples were provided by the indigenous and developing world communities referred to above who were research participants and donors in various research programs—the Chinese, Italian, San, Hagahai, and Havusapai amongst others. The question to be asked is not whether these communities were prepared to contribute their samples for the benefit of humankind. The more pertinent question appears to be whether the call for altruism or solidarity alone makes any sense in view of the overall distribution of research burdens and benefits, and the possibility of exploitation of developing world communities in the research exchange. Whether they were reciprocated commensurately in the exchange, and whether the gains made by the research organisation were inappropriate if not wrongful, need to be answered.

4.4 Summary and Conclusion

In conclusion, this chapter has examined whether a call for altruism, motivated by the notion of solidarity with humankind, provides a cogent reason why benefit sharing should not be considered in the context of the taking of DNA samples from research participants in developing countries. A brief explanation of the concepts of altruism and solidarity provide a glimpse into some of the complexities that surround these terms, and into their relationships with the notion of reciprocity. Chadwick and Berg's argument, to the effect that notions of solidarity and equity provide an ethical duty for participants to donate genetic samples rather than a right to expect benefits as a mechanism to ensure fairness and reciprocity, was tested by examining the practice of donations of human DNA for bio-banks and other research purposes in the affluent world, where altruism appears to be the prevailing ethos.

The most powerful argument against the call for altruism was found to reside in the undeniable fact that adequate recompense in the form of an operational health care system does not exist in the developing world. The *fair exchange model* described by Schroeder and Gefenas is simply inapplicable in the developing world, where tissue donors cannot rely upon the benefits provided by a functioning health care system. Without any form of compensating advantage in the form of health care benefits, requests that humans donate DNA freely in the developing world run the risk of ignoring the potential exploitation inherent in that exchange. Cases such as the Tristan da Cunha islanders support the argument that where the research enterprise stands to benefit materially from donated human DNA samples, calls for altruism and solidarity on the part of tissue donors, unreciprocated in any way, simply lack the element or perspective of fairness. The defining characteristic of Mayer's second class of exploitation, namely insufficient benefit, would be present in such a situation. It is therefore concluded that the call for altruism cannot succeed in the face of the need to consider some form of reciprocal benefits in order to ameliorate the exploitation inherent in accessing DNA samples from developing world communities.

References

Berg, K. (2001). The ethics of benefit sharing. *Clinical Genetics, 59*, 240–243.

Chadwick, R., & Berg, K. (2001). New ethical frameworks for genetic databases. *Nature Reviews Genetics, 2*(4), 318–321.

Cunningham, H. (1998). Colonial encounters in postcolonial contexts: Patenting indigenous DNA and the human genome diversity project. *Critique of Anthropology, 18*(2), 205–217.

De Beer, P., & Foster, K. (2008). *Sticking together or falling apart? Solidarity in an era of individualization and globalization.* Amsterdam: University Press Amsterdam.

Herman, B. (1993). *The practice of moral judgment.* Cambridge: Cambridge University Press.

Johnsson, L., Helgesson, G., Rafnar, T., Halldorsdottir, I., Chia, K. S., Eriksson, S., & Hansson, M. G. (2010). Hypothetical and factual willingness to participate in bio-bank research. *European Journal of Human Genetics, 18*, 1261–1264.

Kant, I. (1797) (1996). Metaphysics of morals. In M. J. Gregor (Ed. and Trans.) *Practical philosophy. Cambridge edition of the works of Emmanuel Kant.* Cambridge: Cambridge University Press.

Kettis, A., Ring, L., Viberth, E., & Hansson, M. G. (2006). Genetic research and donation of tissue samples to Biobanks. What do potential sample donors in the Swedish general public think?. *European Journal of Public Health, 16*, 433–440.

Knoppers, B. M. (2000). Genetic benefit sharing. *Science, 290*(5489), 49.

Mauss, M. (1950). *The Gift first published 1950*, English edition by Rutledge Classics 2002, Oxford, England.

Mayer, R. (2007). What's wrong with exploitation? *Journal of Applied Philosophy, 24*(2).

Pensky, M. (2009). *The ends of solidarity: Discourse theory in ethics and politics*. Albany: State University of New York Press.

Schroeder, D., & Gefenas, E. (2010). Vulnerability: Too vague and too broad. *Cambridge Quarterly for Healthcare Ethics, 18*, 113–121.

Schroeder, D., & Pogge, T. (2009). Justice and the convention on biological diversity. *Ethics and International Affairs, 23*, 265–278.

Scott, N., & Seglow, J. (2007). *Altruism*. Maidenham: Open University Press.

Simm, K. (2005). Benefit sharing: An enquiry regarding the meaning and limits of the concept in human genetic research. *Genomics, Society and Policy, 1*(2), 29–40.

Smith, A. (1776) (1976 edition). *An enquiry into the nature and causes of the wealth of nations* (R. H. Campbell & A. S. Skinner, Edited). Oxford: Clarendon Press.

Titmuss. (1973). *The gift relationship: From human blood to social policy*. New York: Vintage Books.

Titmuss. (1997). *The gift relationship: From human blood to social policy, expanded and updated*. In A. Oakley & J. Ashton (Eds.). New York: New Press.

Chapter 5
The 'No Value Added' Argument

Abstract The third argument against benefit sharing in the context of human genetic resources is based upon the notion that humans, unlike researchers, not having done any essential task or act of competence in relation to their DNA, should not be entitled to any form of benefits. Two arguments are employed to counter this claim. Firstly an economic argument describes how the laws of supply and demand create the value of products, such as human DNA. As the value of such a product increases, those lawfully in possession are entitled to dispose of it and to receive its value without being required to do any essential task. The second argument draws an analogue between indigenous peoples with lawful possession of mining lands, who are entitled as of right to withhold consent to prospect and to mine the lands. This right is similar to that of individuals and communities to withhold, or alternatively to allow, research use of their DNA. Benefit sharing is shown to be essential in the mining context, despite the fact that the resources available to the indigenous peoples through their land have not been produced or added to by them. The conclusion is arrived at from both arguments that there is no requirement for individuals or research communities to have added value to their genetic resources in order to justify the need for benefit sharing.

Keywords No value added to resources · Bioprospecting · Indigenous peoples mining rights · Rights to use DNA

The *no value added* argument presented by Berg against the right for individuals to receive benefits holds that it is not easy to find specific arguments for such a right for the reason that the individuals themselves, "have not performed any act of competence to create DNA" (Berg 2001, p. 242). Whilst admitting that provision of samples for genomic research can lead to significant revenue, Berg's argument is that individual donors have not done anything to add value to their samples. He explains further,

> [...] thus they cannot have a right similar to that based on intellectual property or patent acts. Their body has simply synthesized a compound based upon hereditary instructions

R.S. Chennells, *Equitable Access to Human Biological Resources in Developing Countries*, DOI 10.1007/978-3-319-19725-8_5

> passed on from the parents. If a person's DNA becomes valuable, it would be because of something that researchers have done with it or because of some pre-existing knowledge attained by the world of other scientists at an earlier stage (ibid p. 242).

According to Berg, individual donors, having done nothing, no act of competence or essential task, do not deserve any benefits. There is therefore, no moral or other reason why they should be afforded any benefit in exchange for their DNA samples.

As suggested by Chokshi, there are at least five different groups who could make a claim for benefits, for example in the case of the discovery of an antimalarial molecule. These would include the subjects themselves, the health professionals who diagnosed and treated them, the epidemiologists who managed the study, the geneticists who produced the result, and the company that makes the end product (Chokshi et al. 2005, p. 11). The question arises then if the subjects should receive benefits, in the light of all of the other contributions which do require competence and performance of essential tasks?

Berg's proposition of the *no value added* problem will be countered with two arguments intended to show that communities have the right to benefit from their DNA[1] in circumstances where such DNA has value, which is not caused or provided primarily as the result of any effort on their part. Firstly, by explaining how supply and demand operate in economics generally, assuming the legality of the exchange and secondly, by reference to the broad analogy of communities who acquire wealth in circumstances where the prospecting or discovery process is carried out by external parties.

5.1 An Economic Argument

In economics, the value of any commodity is primarily influenced by, and is set by the relationship between the supply and the demand (Besanko and Braeutigam 2005). The laws of supply and demand are often graphically represented, expressing the dynamic relationship between the two, and showing if demand for a product increases, the price rises, and vice versa. To elaborate this point, the four basic laws of supply and demand according to Besanko and Braeutigam are, (1) If demand increases and supply remains unchanged, a shortage occurs, leading to a higher equilibrium price. (2) If demand decreases and supply remains unchanged, a surplus occurs, leading to a lower equilibrium price. (3) If demand remains unchanged and supply increases, a surplus occurs, leading to a lower equilibrium price, and (4) If demand remains unchanged and supply decreases, a shortage occurs, leading to a higher equilibrium price.

[1]There may be other reasons why individuals should not benefit from their DNA, for example, the general prohibition against commodification of the human body. Here I will only deal with the *no value added* argument.

Demand for products is however not necessarily created by the holder of a commodity, and can change dramatically due to extraneous events such as changes in fashion or advances in technology. The holder is not required to do anything in particular in order to be fully entitled to the market price. For example, Ostriches native to Africa were for many years harvested primarily for their meat, until the world of European fashion decreed the feathers a desirable commodity, setting off an insatiable demand for the product. The Kiwi bird from New Zealand was similarly only kept for its meat, until it was discovered that Kiwi oil extracted from the stomach of the bird is a highly prized beauty product.

The values of gold, diamonds and other minerals are continuously in fluctuation, influenced by fashion as well as technology and other global factors. The values of products such as these are primarily determined not only by their inherent component utility or anybody's effort but also by extraneous events influencing the onset of human demand in the marketplace. The essential point is that the owner, custodian or lawful possessor of such products automatically benefits from the increase in demand, and is entitled to negotiate a commensurate price, without having necessarily played any active role in increasing the market value. The mere fact of lawful possession associated with the right to refuse to or to dispose of the item is sufficient to ground her entitlement to the full amount of the value. This begs the question then as to the situation regarding human research participants, who possess and have the right to dispose of their DNA. The question being addressed in order to engage with Berg's argument is, are the owners of human tissue required to do any particular act in order to be entitled to benefit from the enhanced value for research purposes of their DNA samples?

Prior to the era of genomic research, there was little demand for human biological materials other than for blood transfusions, forensic analysis and related practical matters. However following relatively recent radical advances in biomedical sciences it has now become a key ingredient in genomic research, where scientific advances coupled with their potential to lead to commercial value have been a rapid and relatively recent occurrence.

Through no specific effort on the part of their carriers, human genetic resources have, as a consequence of the priorities of biomedical research together with the workings of the patent system described above, acquired potential commercial value and have thus in the manner described in chapter three, become commodified (Hayden 2007). Demand for DNA has increased from the fields of pharmacogenomics to population genetics. Consequently and in accordance with the laws of economics the value of human genetic bio-banks, driven ultimately by the requirements of the commercial sector market, has risen accordingly (Womack and Gray 2000, p. 251).

In summary, it would appear to be uncontroversial to state that individuals and groups have possession and control over their own genetic resources, subject to the laws of the land, which resources without any effort on their part have acquired undeniable commercial potential within the field of biomedical research. The demand for human DNA has increased and will probably continue to do so. Potential research participants clearly have the right to either refuse to participate in research, or alternatively to take note of the demand or need for their DNA, and to consider the possibility of receiving commensurate collective benefits in exchange for their participation. They might well have not contributed to the value now acknowledged in this resource, but they have the right to control access to its use, and the right to choose whether to participate or not. It follows that any economic advantage which might arise from their decision whether to participate or not belongs to them alone, without any consideration of whether or not they have in any way added value to their genetic resources.

5.2 The Prospecting Argument

The second argument against the notion that individuals and communities have no right to receive benefits because they have added no value to their DNA is related to and builds to an extent on the preceding argument, but focuses upon the component of prospecting that is inherent in all genetic research. The Oxford dictionary definition of the verb *to prospect* means, "To search for, or to seek mineral deposits, especially by drilling and excavation".[2] A simple analogue is proposed for this argument, with acknowledgement that some aspects of the transaction are intrinsically different. Consider the example of a rural community that owns a tract of land which is suspected to have valuable mineral resources, where the value of the resource is hitherto unknown, and is not able to be established without an externally instigated exploration, or prospecting process.

Indigenous (Aboriginal) communities of rural Australia, for example, own vast tracts of potentially mineral-rich land, and are customarily approached by prospecting and mining companies requesting the rights of access (Altman 2009). The indigenous communities are required to have survived on and protected their land from generation to generation, which is a prerequisite for land ownership and entails intrinsic cultural and other values. This could however barely be described as falling within the meaning of any acts of competence or as adding particular commercial value to the land.

A sceptic might argue that the indigenous communities *own* their land, which is not the case with individuals and their genetic constitutions. It is proposed not to engage with the notion of ownership as opposed to possession or custodianship of a resource; it is sufficient for this argument that both the indigenous community as well as the individual and collective research participants have legal control over the resource in question, which provides the commensurate right to approve or deny access. This ability to choose whether to enable or to refuse access to an outside party is regarded as a sufficient basis for the analogy.

Such an indigenous community would typically lack the capacity to exploit the value of its land resource commercially, or to even ascertain the potential value of

[2] [Online] Available at http://oxforddictionaries.com/definition/english/prospect.

the minerals below the surface of the land. It requires outside intervention from the world of industry and commerce, recognising that the tribe's land might well contain valuable mineral deposits, to investigate and establish the presence of commercial value. In order to ascertain whether deposits of commercial value exist, a mining company is required to obtain permission from the tribe and thereafter to carry out a prospecting process.

The prospecting process entails the exploring, researching and evaluating of the land for the presence of potential mineral wealth. A prospecting right to explore and evaluate is of considerable value in itself for the reason that it is indispensable for achieving the overall objectives of the company, and it is usually linked to the consequent right to exploit any possible leads or discoveries that may emerge. Competing companies customarily bid against one another or place competing tenders for the valuable prospecting rights, after which the landowning community approves the bid of a suitable company, and a prospecting lease or agreement is negotiated. A prospecting agreement then typically awards the right to explore to a particular company, and provides for various negotiated communal benefits such as the development of roads and other facilities, in addition to cash payments. If valuable minerals are found in the prospecting process, a discovery analogous to the detection of a useful active ingredient in biomedical research, the mining company exercises its right to commence with proper mining, in which event a further substantive agreement is entered into with the community dealing with long term mutual rights, duties and benefits associated with the commercialisation process.

In this type of case, the indigenous community, by granting access to the mining company for the purpose of prospecting, and agreeing to the prospecting process of the mining company, might find that their land resource has acquired considerable value in the commercial world. Whilst the providing of access may well entail certain inconveniences, such as light drilling of test holes, the indigenous tribe will have performed no particular act of competence in relation to the emerging value other than legally and physically maintaining the land. After a process of information sharing in relation to the risks and benefits associated with the intended prospecting and mining, analogous to the process leading up to the acquiring of informed consent in the biomedical research arena, the consent of the tribe enables the skilled mining engineers and chemists to apply their various crafts and establish the presence or otherwise of value.

Compare the prospecting and mining example with the case of an indigenous or rural community whose genetic make-up has become of interest to research, and which is approached with a request that its members should participate and donate samples for some form of genomic, or population wide genetic project. The community might similarly have little scientific or commercial capacity, and would have done nothing in particular to enhance the value of the now sought-after and potentially valuable resource under its collective custodianship and control. In the same way as the land-owning Aboriginal group described above, the indigenous community would have scant knowledge about the value of their DNA, other than the fact that it is now being sought after by researchers from the developed world.

It is therefore suggested that such an indigenous community would similarly have the clear right to either refuse to allow access, and if the latter, to negotiate and enter into a simple form of prospecting agreement with the institution of its choice, whereby its members provide the raw material of their DNA to the research group for research and evaluation. Therefore in seeking to reach such an agreement, which would record the community's informed consent in relation to the granting of access, it would similarly have the right to be fully informed about the intended use of the research, the likelihood of commercial utilisation, and the possibility of reaping present or future collective benefits for the community.

To bring the discussion of prospecting for minerals to a close, it is suggested that the process is analogous to and comparable with bio-medical research. The value inherent in DNA is not immediately obvious, or even ascertainable at the time of exchange prior to and without a procedure of scientific exploration as required for mineral prospecting and described above. Despite having done nothing in particular to discover or extract the value, certainly no form of essential or overt contribution of the form envisaged by Berg, the communities described above in both cases—minerals and DNA—have every right and expectation to either refuse to participate, or alternatively to agree to participate upon fair terms which provide for appropriate mutual benefits in the exchange in the event of commensurate success. The latter process seeking agreement correlates with the informed consent process that precedes research. In such cases, the act of competence is the prospecting, which is in all cases carried out by skilled third party specialists. It is therefore sufficient for this argument, that the community holds a resource of indeterminate value, and has the crucial right to provide or not to provide access in a process of exchange, which takes into account the future potential value of the commodity.

In conclusion, this discussion on the *no value added* objection has explored whether individuals or communities are required to do some essential tasks, or otherwise add value in any particular manner, that is to perform an act of competence, in order for them to have a legitimate claim to the value of their genetic resources. The two arguments drawn upon above would suggest not. The basic laws of economics dictate that a community's simple right to approve and provide access to its members' DNA, without any moral or other requirements, provides it with a complete entitlement to receipt of the full value. The mining analogy argued that a community, which is in possession of a resource of an entirely unknown commercial value, possesses an unassailable right to either allow or to refuse access, in the form of prospecting, in order to ascertain the inherent value of the resource.

It is lawful possession, together with the associated right to say 'no' to other parties, which form the essential link and basis of the comparison. These two arguments support the conclusion that a community's failure to have done anything or to have performed a significant act should not detract in any way from its right to ascertain, and to perhaps benefit collectively from, any value that may be associated with or flow from the DNA samples provided by its members for the benefit of biomedical research.

Having addressed arguments against benefit sharing now, it is convenient at this stage to summarise briefly the trajectory of the discussion thus far, and to point towards the way ahead. The book commenced with a description of the problems within genomic research that motivated the overall topic, and which led to proposals for benefit sharing in the research exchange of human DNA following the example set by the CBD. One of the purposes of research ethics guidelines being to avoid all forms of exploitation in biomedicine, the meaning and nature of the word exploitation was then explored in order to provide guidance for the further discussions. Thereafter, the book commenced with an investigation of the first research question, which was whether and in what circumstances benefit sharing is appropriate for human genetic research. This required a detailed analysis of the three most significant objections to benefit sharing, namely the *common heritage of humankind*, the *altruism*, and the *no value added* arguments. These three arguments against benefit sharing were in each case refuted, primarily by examining the unfairness in the exchange of human DNA where benefit sharing is not considered. Applying Mayer's second class of exploitation, which is concerned with fair requital during a transaction, the mechanism of benefit sharing was found to be necessary in order to ensure justice.

The second research question addressed in this book encompasses the final, and possibly the most persistent obstacle in the way of benefit sharing, namely the objection that any benefit-sharing offered to research participants might fall foul of the prohibition against undue inducement or coercion. These latter two concepts are closely related, with their roots and meaning firmly sourced in the law relating to contracts. Noting that the overall purpose of this book is to examine equitable access to human biological resources, words which could be equated with the phrases *just access* or *fair access* during exchanges of human genetic resources for research, one could equally state the overall purpose as being to bring about justice in this bilateral exchange between researcher and research participant community. Prior to examining the meaning of undue inducement and coercion, the following chapter will therefore explore the origin and nature of these two legal doctrines that were developed to combat exploitation from within the broad realm of justice.

References

Altman, J. (2009). Benefit sharing is no solution to development: experiences from mining on aboriginal land in australia. In R. Wynberg, D. Schroeder, & R. Chennells (Eds.), *Indigenous peoples, consent and benefit sharing: lessons from the san hoodia case*. London: Springer.

Berg, K. (2001). The ethics of benefit sharing. *Clinical Genetics, 59*, 240–243.

Besanko, D., & Braeutigam, R. (2005). *Microeconomics*. London: Wiley.

Chokshi, D. A., & Kwiatkowski, P. (2005). Ethical challenges of genomic epidemiology in developing countries. *Genomics, Society and Policy, 1*(1), 1–15.

Hayden, C. (2007). Taking as giving: Bioscience, exchange, and the politics of benefit sharing. *Social Studies of Science, 37*, 729–758.

Womack, C., & Gray, N. M. (2000). Human research tissue banks in the UK: National health service: Laws, ethics, controls and constraints. *British Journal of Biomedical Science, 57*, 250–253.

Chapter 6
Justice and Exploitation in Bilateral Exchanges

Abstract This chapter contributes to the final argument against benefit sharing, namely that coercion or undue inducement would be a likely and negative outcome. The broad concepts of global justice that frame the area of debate are first traversed, in order to examine how justice and law have approached potentially unfair (or illegal) exchanges in the past. After a discussion of global justice and international law, domestic law is analyzed by means of the conceptual framework provided by Thomas Pogge. Four different levels of justice are outlined, each of which is divided into a procedural and a substantive component, and where the final level includes distributive, commutative, restitutive and retributive justice. The research exchange of genetic resources is conceptually placed within commutative justice, a form of corrective justice originally distinguished by Aristotle who usefully analyzed law and injustice between parties. An examination of subsequent responses to injustice between parties leads to the ancient law of equity that originated in England, and which contributed significantly to improving the fairness of modern contract law. Finally the three distinct legal actions in contract law which originated in the law of equity, namely coercion, undue inducement, and unconscionable dealing, are discussed, offering as they do clear criteria and guidelines for assessment of unfairness in differing circumstances. These criteria provide an analytical framework for the next chapter's understanding of undue inducement and coercion in the context of benefit sharing for human genetic resources.

Keywords Justice · Contractual exploitation · Undue inducement · Coercion · Unconscionable dealing · Commutative justice · Contract law · Equity law

The final impediment to benefit sharing now stands to be addressed, namely the concern in bioethics that such benefits might coerce or unduly influence vulnerable research participants into providing their genetic resources for research, against their interests and without due consideration of the potentially harmful consequences. This objection states in effect that the benefit-sharing proposed in order

R.S. Chennells, *Equitable Access to Human Biological Resources in Developing Countries*, DOI 10.1007/978-3-319-19725-8_6

to ensure justice in the research exchange is inappropriate, as it can under certain circumstances exploit vulnerable research participants by coercing or inducing them to accept inappropriate risks.

The form of unfairness or exploitation being examined here relates to the manner or procedure in which consent to a transaction is brought about, rather than to the substantive or outcome-related aspects. The legal framework for such an enquiry is situated within the law of contracts, which governs all voluntary transaction, and which in turn belongs within the broader realm of justice. Justice will be briefly examined in order to understand and inform this analysis of procedural exploitation within exchange transactions.

The broad concepts of justice and then of global justice will first be briefly discussed, in order to provide the context for the striving for fairness that informs international and national legal systems. Justice and domestic law will be introduced by Thomas Pogge's conceptual framework for justice, which will describe and situate the most relevant concepts for this book, including procedural and substantive justice. Thereafter, the Aristotelian origin and meaning of the two forms of justice relevant in bioethics, namely commutative justice—or justice in exchange—and distributive justice will be described, after which the discussion will turn to modern domestic law. A particular emphasis will be placed on the ancient law of equity which evolved out of a perceived need for greater justice in the law, and which infuses modern legal defences against interpersonal exploitation.

Following this the discussion will turn to the law of contracts which governs all voluntary human exchanges such as consent forms and research agreements, with a final focus on the three forms of contractual exploitation, namely unconscionable dealing, duress (or coercion) and undue influence. These three legal doctrines all have their origin in the law of equity, and respond to concerns that mirror the endeavours of the ethical guidelines, namely to prevent exploitation of the vulnerable. The overall purpose of the chapter is to provide a clear understanding of the characteristics of contractual exploitation as understood by the law, in order to inform the discussion on undue influence and coercion that will follow in the next chapter.

6.1 The Concept of Justice

The famous question asked by Socrates in Plato's Republic, namely *what is justice?* has been debated by philosophers for millennia. In ancient Greece two words were used that might be translated into English as justice, the first being *dikaiosune*, and the second being *ison*, the former meaning righteousness and the latter word translated more accurately as equality (Vlastos 1962). On a personal level justice is to be aspired after as both Plato's "summary virtue" for citizens, where "doing one's own" is required by all members of a community, and where equality, desert and fair exchange are central themes in the discourse. Parallel legal philosophies emerged over millennia from the cosmologies of the East based upon concepts of equilibrium and harmony, focusing on natural justice and its political implications revealing the pervasiveness of humankind's quest for justice (Chun 2010, p. 59).

The durable metaphor of Lady Justice, a blindfolded woman holding a human scale in one hand and a sword in the other, illustrates the impartiality inherent in the concept. The scales represent the competing claims to be weighed; the blindfold indicates visible lack of bias, and the sword the coercive power of the judge or supreme power (Luban 2001, p. 23). On a personal level, the golden rule ascribed to various sources including the Bible namely to "do to others as you would have them do unto you", developing in similar forms from Confucius through all of the major religions and ethical traditions (Kidder 2003, p. 159), is a reminder of the everyday human touchstone that underlies the concept of justice and fairness. This rule or concept represents in the words of Ken Binmore a "fairness program within us that runs well below the level of consciousness" (Binmore 2006, p. 4).

Human rights as sources of justice have evolved during the past half century into a substantial set of laws and standards with pervasive jurisdiction in international law, providing an example of how the realisation of justice continues to develop and progress with time. Having laws in place is however no guarantee of justice, and whilst positivists believe the law is the law and should be strictly so applied (Weinreb 1987, p. 193), others of a liberal bent seek greater flexibility in the interests of justice. It is not the realm of this book to attempt to describe further all the extensive contributions towards justice, from utilitarians and consequentialists to those who base their analysis upon rights theories or some form of social contract (Halgarth 1998). The latter, which included Locke, Rousseau and Rawls, also drew on natural rights, and drew inspiration from Aristotelianism, Roman law and Stoic views on the state of humankind in nature (Cooney 1998). The social contract provided a theory of institution forming that is reflected by the way international law and constitutions, including the laws governing the domain of biomedical research, developed today. According to David Gauthier (1977), the idea of a social contract is now so engrained in modern consciousness as to constitute more than a modern ideology, providing a rationale for the way we behave towards each other. As he states, "the justification of rights and duties, institutions and practices, is to be found by regarding them as if they were contractual, and showing the rationality of this hypothetical contractual base" (ibid, p. 115).

6.2 Global Justice and International Law

Global justice aspires after justice in international affairs, including the distribution of rights, responsibilities and flourishings not only of states but also of peoples and citizens. International law has evolved in a largely contractual manner in order to regulate these aspirations and dealings, now divided into numerous categories such as the law of treaties and contracts, international tort or criminal law, international resource law, human rights law, and many more (Franck 1995, p. 5). International law has been defined as, "a body of rules regulating the relations of states but also the relations of those other legal entities which are recognized as possessing international personality at any given time" (Wallace 2005, p. 2).

Many of the principles of justice that embody international law have ancient origins, such as the *jus cogens* (the law according to thinking persons) which refers to norms from which no derogation is permissible, such as against genocide, racial discrimination, slavery, piracy and which all states owe to the international community (Brownlie 2003, p. 515). Prior to the Second World War international law comprised solely of relationships between nations, but since then other legal entities including multinational corporations, non-government organisations—such as the International Committee of the Red Cross and Amnesty International—indigenous peoples and even individuals have acquired certain forms of international personality under specific circumstances, and are thus to an extent included (Dugard 1994, p. 2).

International law is founded essentially on consensus due to the lack of a higher authority, which is determined firstly by the practice of states known as customary international law and secondly, by agreements, treaties or conventions (Wallace 2005, p. 5). Consensus in this arena, as described in Sect. 39 of the International Court of Justice, is drawn from various conventions, customs, general practice and the like. This consensus is forged either out of necessity, or out of the need of states and their citizens to co-exist and respond to changing circumstances.[1] For example, after 1945 the enormity of the atrocities committed by the Nazi regime led to a revocation of the prohibition on intervention in domestic affairs of states enshrined in Sect. 15(8) of the Covenant of the League of Nations, and resulted in the United Nations Charter (UN 1945). Article 1 of this Charter includes amongst the many purposes of the United Nations the "promotion and encouragement of human rights", and article 56 contains a pledge by all nations to take "joint and several action" to achieve these purposes (Dugard 1994, p. 199).

The previously unchallenged right of states to the privacy of their domestic affairs thus became a thing of the past. The human rights revolution continued with the Nuremberg trials, followed by the United Nations Declaration of Human Rights of 1948 which laid a framework of principles consisting of broadly shared moral and ethical precepts. Many of the rights that are now regarded as part of customary international law were at that early stage optimistic or aspirational in their reach, and no

[1]Section 38 of the statute of the International Court of Justice defines the international laws to which it shall apply as being (a) international conventions whether general or particular, establishing rules expressly recognized by the contesting states, (b) international custom, as evidence of a general practice accepted as law, (c) the general principles of law recognized by civilized nations, and (d) judicial decisions and the teachings of the most highly qualified publicists of the various nations, as subsidiary means of the determination of the rule of law. [Online] Available at http://www.icj.org.

enforcement mechanism existed to ensure compliance by states. Nevertheless the declaration inspired several increasingly stronger conventions,[2] regional conventions and national Bills of Rights, which created a new human rights paradigm for international law (Dugard 1994, p. 204).

The relevance of the above lies in the fact that all the conventions, declarations, laws and guidelines governing genetic resources and biomedical research are similarly evolving in a paradigm of continually intersecting influences of perceived law and morality (Boylan 2011, p. 50). Most importantly, research, debates and proposals that attract sufficient support contribute towards and result in the development and evolution of such international law. As stated by one commentator, the broad principles of social justice that have been promulgated as 'law' in the arena of international trade of genetic resources generally, as laid down in the CBD and the Nagoya Protocol regarding access and benefit sharing, will increasingly intersect with, inspire and in turn be influenced by evolving norms governing biomedical research (Faunce 2005).

6.3 Justice and Domestic Law

The broad overview of the concept of justice sketched above provides a framework for understanding voluntary human transactions of all varieties, including genomic research. In order to understand how justice is translated into practical or domestic law, this section will first describe Thomas Pogge's (2006) theory of justice, which provides a framework for the discussions that will follow. Then procedural and substantive justice will be discussed, followed by Aristotle's division of law into distributive and corrective justice. These will introduce the concept of the interpersonal forms of law that guide and limit the genomic research being discussed in this book, leading to an analysis of three forms of exploitation addressed in the law of contract.

6.3.1 Thomas Pogge's Conceptual Analysis of Justice

A single inclusive theory of justice remains elusive and perhaps not even necessary, supportive of Aristotle's conclusion that to seek such a unified concept of justice is futile (Winthrop 1978, p. 1201). However, a modern analysis of justice is useful to frame and locate this enquiry into the fairness of the exchange of human genetic resources. Thomas Pogge, in a treatise on Justice, provided such an analysis by first describing the four domains of application or *judicanda*[3] against which

[2]For example, the International Covenant on Civil and Political Rights, and the International Covenant on Economic, Social and Cultural Rights, both of 1966.

[3]From the Latin *judicandum*, that is that which is to be judged.

justice can be evaluated, namely (a) individuals and collective actors, (b) the conduct of such actors, (c) the social rules, laws, institutions and conventions applicable, and (d) the actual outcomes or states of affairs being observed (Pogge 2006, pp. 8–11). The first two *judicanda* namely the parties and their conduct have a certain primacy in that they influence both the social rules and eventual outcomes (ibid, p. 865). Above and overlaying these *judicanda,* each of which can be separately assessed and evaluated in a justice enquiry, such as a particular allegation by one party of exploitation in an exchange, Pogge then conceptually poses three dimensions or distinctions of justice. The first dimension is between *first-order justice* and *procedural justice*, the former describing the particular allocation of benefits and burdens, also termed *substantive justice* in jurisprudence, whilst *procedural justice* on the other hand is the formal procedure or manner in which the outcome was determined. For example, an unequal division of spoils between two partners is first order or *substantive* injustice, whereas the fact that one party decided on the allocation without consulting the other would comprise the *procedural* unfairness. Procedure and substance are once again inextricably linked in justice.

Pogge's second dimension of justice, depicted as an enquiry overlaying and including the first dimension, is comprised of formal and material justice. Formal justice is the prevailing legal system, in which consistency of judgement is provided in the system of precedent; one that ensures similar treatment in former decided cases, that establishes clear principles and laws, which engenders confidence in its fairness, and that provides the metaphorical blindfold over Lady Justice. Material justice on the other hand, is again comprised of the substantive or material outcome of a case. Whether taxpayers are being fairly treated, whether certain classes of individuals are unfairly discriminated against, whether certain people or classes are short changed, exploited or aggrieved.

The third dimension of justice envisages the investigation into the particular circumstances of material justice in each case into four distinct categories or domains, namely distributive, commutative, restitutive and retributive Justice. Distributive justice analyses the division of resources between people, groups or even nations, whilst commutative justice deals with the equivalence of value in exchanges between two or more parties. Restitutive justice deals with restitution of violations of rules or laws, modern law of tort, whilst retributive justice deals with punishment for such violations (ibid, pp. 8–11). It should be noted that some scholars refer to the latter three under the collective rubric of *corrective justice*, in that they all deal with injustice inflicted by one person on another, which stands to be corrected (Weinrib 2002, p. 349).

Distributive justice on the other hand is not triggered by or dependent upon a wrong committed by one party on another. In each of these four categories or domains of justice, all of the foregoing dimensions are examined, and both procedural and substantive aspects apply. The procedure consists of the rules governing the processes and functioning of law, and the substance consists of the factual outcomes affected by law and guidelines affecting the allocations of benefits and burdens. Both the procedural and the substantive aspects of any one case may be found to be unjust for formal reasons or for material outcome-evident reasons, visible in the actual outcomes experienced by the parties (ibid, p. 10). In total, Pogge's model identifies sixteen conceptual boxes for assessment of justice concerns, each of which can operate across all of the four *judicanda* described at the beginning of the analysis. The complexity of the composite conceptual framework is indicated by the fact that no less than 64 separate categories can be identified and assessed.

Using this conceptual structure to examine and frame the Tristan da Cunha case for example, we start by placing the parties, their conduct, the prevailing laws and the final outcomes of the case within the *judicanda* that are to be analysed. The first dimension enquiry assesses to what degree first order justice pertains in the division of benefits between the parties, and whether the procedures, not just between the islanders and the researchers but also regarding all external institutions, were fair and just. The second dimension of enquiry takes into account the formal system of justice applicable to the transaction, whether the company and islanders complied with existing laws, and whether other role players—ethics committees, downstream research partners—complied with existing laws or standards.

The material justice enquiry at this level explores the material aspects of the final outcome, the distribution of benefits, and would include perceptions of exploitation or injustice. The third dimension finally expands this enquiry into the four systems of justice described above, each of which has its own laws, rules and systems on both procedure and substance. At this third dimension of enquiry, the Tristan da Cunha case could, for example, be analysed both as a matter of distributive justice, in that at a broad level it concerns the distribution of a common property resource for global research with potential impact across the world, and also as a matter of commutative justice in that the case began with an exchange between two primary parties, where justice in that particular exchange is most relevant. This cursory application of Pogge's justice framework illustrates the sheer complexity of attempts to depict a unified theory of justice, and again emphasises the pervasive distinction between substantive justice on the one hand, in other words the perceived justice of the outcomes, and procedural justice on the other.

6.3.2 Procedural and Substantive Justice

These two components are present in every aspect of a justice enquiry (Franck 1995, p. 6). The first aspect often described as *due process* provides good order and legitimacy, entailing rules of process rooted in a formal framework of how rules are to be made, interpreted and applied (ibid, p. 7), whilst the second involves the consequential material effects of the particular law under scrutiny, namely the application of both corrective and/or distributive justice (ibid, p. 8). Both procedural and substantive justice are relevant for example in an institutional process such as the evaluation by a REC, and need to be satisfied in order for any outcome to be deemed fair.

Procedural justice, at once an inherent part of every system of justice as well as a field in its own right, has roots deep in the law of equity, and is an absolute requirement for any perception of fairness regarding a distribution of resources (Tyler 1987). Empirical research on procedural justice in practice has supported the conclusion that the fairer the procedure used to determine outcomes, the more psychologically acceptable the outcomes will be (Folger et al. 1984). Both Aristotle and John Rawls believed that without a fair institutional system, which creates fair laws, no justice could exist.

Most constitutions expressly affirm for citizens the right to procedural fairness or due process, also defined correspondingly in reverse as the duty of the authorities to act fairly, based upon the principles of natural justice (ibid, p. 188). Procedural justice is then achieved when the rules and procedures brought to bear and which will determine matters of significance are infused with principles of natural justice, which ensure that all information, all sides of the situation are weighed up in a just process, fairly and without bias. Substantive justice on the other hand examines the material manifestation or distribution of benefits and burdens that affect all parties to the enquiry (Deutsch 1985). Situations involving both distributive justice as well as justice in exchange result in substantive or material outcomes, each of which are examined by one or other of the two forms of justice.

6.3.3 Aristotle's Distributive and Corrective Justice

Aristotle's theory of justice is a good place to begin a brief exploration of the distributive and corrective justice of Pogge's framework, bearing in mind this book' preoccupation with commutative justice or justice in exchange. His *Nicomachean Ethics*[4] was intended to be a comprehensive investigation into the end or purpose of all human action, which he regarded as *eudaemonia*, namely the point of life, the human good, or happiness (Aristotle 1094a, pp. 1–3). Of all the virtues dealt with in the *Ethics*, justice was described as the last of the virtues of character. Whilst he regarded "general justice" as being complete virtue, or "whatever produces and maintains happiness and its parts for a political community" (Aristotle 322 (1985), 5.21, 5.23), he presented a theory of "particular justice" which is divided into two forms, namely distributive and corrective justice (Aristotle, 1130b, 30–1131a, 1).

Distributive justice as the first of Aristotle's two forms of justice is concerned with, "the distribution of honours or wealth or anything else that can be divided amongst members of a community that share a political system" (Aristotle, 322 (1985), 5.34). Describing the adjudication process Aristotle called this allocation

[4]The Nicomachean Ethics will be referred to as "Aristotle", followed by the Bekker numbers named after the German Scholar Immanuel Bekker who published the most authoritative edition of Aristotles works in Greek. Where other references are used this is indicated.

of proportion a geometrical exercise, stating that, "what is just is proportionate and what is unjust is counter-proportionate" (ibid, 5.43). According to Weinrib (2002), Aristotle means by these words a proportional equality, in which all participants in the distribution receive their shares according to their respective merits under the criterion in question (ibid, p. 349).

In this evaluation of transactions involving research benefits and duties flowing from the exchange of DNA across continents, ultimately the system of rules and principles adopted by the various authoritative bodies will guide the final distributive outcomes. For example, the negotiations between states regarding the pandemic influenza preparedness (PIP) system, which resulted in a new procedural and legal dispensation for all states both providers and users of viruses following the withholding of samples by Indonesia, were described by Wilke as an exercise in distributive justice (Wilke 2012). In other words the agreement provided rules that ensured legitimate dealing and fair distribution of various commodities and benefits—virus samples, rights and duties—in a multilateral system accepted by all participant states. Within this overall system, individual bilateral transactions between parties would include aspects of justice in exchange.

Bearing in mind Rawl's assessment following contractarians before him that justice can only take place within a consensually adopted institution, and as described in the procedural justice section above, clearly systems, procedures and laws are required to be agreed upon in order to give effect to the demand for distributive fairness. The mechanisms envisaged both by the CBD and Nagoya Protocol in respect of non-human genetic resources as well as the PIP agreement and the bioethics guidelines described above thus pursue a distributive justice agenda, setting laws, rules, standards and principles that aspire to bind stakeholders; ensuring a particular form of global justice in the distribution of the burdens and benefits associated with the utilisation of genetic resources in research.

Corrective justice, as the second of Aristotle's forms of justice, concerns bilateral injustice between two parties, what Aristotle termed, "rectification in transactions" (Aristotle, 322 (1985) 5.34). Corrective justice is according to many central to contemporary theories of private law (Weinrib 2002, p. 349), and is the overall term for the three subsidiary types known as commutative justice, restitutive justice and retributive justice. Weinrib describes Aristotle's three subsidiary types of corrective justice as being essentially transactional justice, during which two bilaterally correlated parties are restored to the notional equality that they possessed before the disputed transaction (ibid, p. 349). These were in addition described by Aristotle as being distinguished in two ways, namely being voluntary and involuntary (Aristotle, 322 (1985) 5.51) The category voluntary concerns exchanges between parties, namely commutative justice which is now contained in the field of contract law, whilst involuntary consists of wrongs performed on individuals without their consent such as tort, theft, adultery and assault (ibid, p. 5.52), which occur within the domain of a legal system enforced by a state.

Commutative justice or justice in exchange, the form of voluntary justice most relevant for this book, was according to Aristotle assessed regardless of the morals or intentions of the parties, and focussed purely on the transaction at hand. "Hence the judge tries to restore this unjust situation to equality, since it is unequal" (ibid, p. 5.52). By illustrating what he had in mind as a fair voluntary exchange when parties engage in a transaction permitted by law, Aristotle continued, "for having more than one's own share is called making a profit, and having less than what one had at the beginning is called suffering a loss … and when people get neither more nor less, but precisely what belongs to them, they say they have their own share, and make neither a loss nor a profit" (ibid, p. 5.52). Finally, addressing in a practical manner the focus of countless lawsuits that would be aimed at establishing the damages or requital to rectify the imbalance, he explained that all items in an exchange should be comparable in some way, requiring a "proportionate reciprocity". Aristotle believed that this, his proportion could be calculated with money, in which regard he enquired, "how many shoes are equal to a house" (ibid, p. 5.64).

To apply Aristotle's analysis of justice, reciprocal exchanges such as the provision of human DNA for international genetic research, the subject of this book, involve principles of what he termed distributive as well as commutative justice principles. Distributive justice in biomedical research projects is concerned with the fair distribution of benefits and burdens between all stakeholders involved in the entire endeavour, including those not involved in the original exchange—the public, research partners, other states and institutions—seeking what Weinrib refers to as some form of proportional equality, which takes place by means of the laws, guidelines, contracts and policies that are applicable.

Commutative justice, by contrast, as part of Aristotle's overall system of corrective justice, is concerned with the bipolar exchange between two parties involved in a voluntary exchange, and features the restoration of a notional form of equality or a proportional requital for the original state with which the two parties entered the transaction. The latter is the domain within which exploitation between two parties, as defined by Mayer, takes place. Finally and in regard to both of these very different arenas of justice, Aristotle expressed the belief that equity was applicable in order to supplement formal law and to achieve justice.

6.4 The Law of Equity

The word equity has ancient roots, with its origin in the Latin word *aequitas* meaning just, impartial or fair. Aristotle used the word to mean, "something which, although different from legal justice, is another sort of justice, not generically different from it" (Aristotle 1137a, p. 33–34). Equity developed in the English common law as a set of legal principles separate from and supplementing the strict rules of the common law, the principles of which were then exported in some form to much of the Western world (Worthington 2006, p. 6). In everyday terms, equity is said to "mitigate the rigor of the common law" (ibid, p. 7) and was introduced precisely to deal with the sort of unfairness between parties described earlier, and to ensure the presence of justice in the legal system. According to Hanbury's Modern Equity, the origin of equity was said to be "in justice, beyond human control" and "older than any of its characteristics" (Hanbury 1981, p. 5), both of which descriptions seek to express the inherent and ancient sources of the notion. The Judges of early England, educated in pure law, administered the law of the realm, but when judgments were regarded as unfair, litigants were entitled to appeal directly against unfair judgments to the King who as the supreme sovereign was regarded as the "font of justice" (ibid, p. 8). The King delegated such petitions for justice to the High Chancellor, regarded as "keeper of the royal conscience" and an important member of the King's Council.

The Chancellors were scholars of the classics well versed In Latin and French as well as in classical Roman law, which heavily influenced their rulings and the development of equitable principles designed to bring about justice (ibid, p. 10). Soon the Chancery began to resemble a judicial body, even though it was not a law court as such, and it became known as the "Court of Chancery". By the 15th Century, the judicial power of equity law was firmly recognized, and by the end of the 17th Century only lawyers were appointed to the Chancery (ibid, p. 11). Equitable doctrines became codified as part of the law, and equity law began to rival and conflict with the common law, leading eventually to the *Judicature Act* of 1870 which fused the courts of equity and the common law, whilst their rules and principles remained separate. The Judicature Act 1873 contained a general residual clause, in Sect. 25 which sums up the relationship between equity and common law generally, "generally, in all matters not hereinbefore particularly mentioned in which there is any conflict or variance between the rules of equity and the rules of common law with reference to the same matter, the rules of equity shall prevail" (Hanbury 1981, p. 14).

As colonies gained independence from England and their legal systems began the inevitable process of drifting from the original blend of common law and equity, the rules of equity have nevertheless remained strong, whether codified into the laws of the country or continuing to exist as separate legal principles. For example, in India the common law doctrine of equity had traditionally been followed even after it became independent in 1947. In 1963, the *Specific Relief Act* was passed by Parliament in which most equitable legal concepts were codified and made statutory rights, thereby ending the discretionary role of the courts to grant equitable relief and making equity part of the law (Lewis and Lobban 2003, p. 306).

The main difference between the law of equity and formal law is equity's flexibility, its ability to fashion a specific form of remedy or relief in a case where the common law does not assist. Equity law was primarily concerned with unconscionable behavior, fraud, or untoward behavior, which tended to arise in mistake cases where one party is manipulated or tricked in a dubious manner (ibid, p. 308). One is reminded of Mayer's (2007) categories of exploitation where he referred to dealings that had a certain cheapness or lacked authenticity in the exchange. Whilst the formal law of contract is limited in its response to a wrong, for example, being only able to award damages to a wronged individual, equity provides a more flexible remedy in those cases where simple contractual damages would not be sufficient to heal the wrong.

Equity law is thus able to apportion damages in accordance with the degree of wrongdoing, to take into account the actions and motives of other parties, and to provide new forms of injunction to ensure justice. The court of equity even had the power to formulate new rules where none existed (ibid, p. 310). Lewis and Lobban provide, as an example, of an equitable injunction a situation where a farmer's special cow is stolen by a neighbor. Contract law would only be able to award him damages, calculated in the amount of the value of a cow, which would not satisfy the farmer for the reason that the cow in question was special and normal damages would not do him justice: equity was however able to order the return of that particular cow (ibid, p. 100).

Lord Kames wrote his influential *Principles of Equity* in 1760, in which he recapitulated the theory of moral reasoning, and developed an influential jurisprudence based on equity to be used for the evolution and advancement of law in changing circumstance. He argued persuasively against positivist and utilitarian views of the law, stating that utility in society could never trump justice (Lewis and Lobban 2003, p. 102). In legal systems that follow the English common law tradition, as well as in civil legal systems, the law of equity remains very much part of the law, whilst in continental legal systems, equity has been codified into the common law and judges are afforded greater liberty of interpretation (Ross 1958, p. 284). A number of maxims or principles of equity have become fully incorporated in law, including the well-known *audi alteram partem* (hear the other side) and *nemo in sua causa iudex* (no one may judge his own case), which underpin the essence of fair procedures (Hoexter 2002, p. 191), as noted earlier.

Further comforting maxims for aggrieved litigants are *ubi ius ibi remedium*, meaning where there is a right, there is a remedy, and *ex turpi causa non oritor actio*, meaning no person may institute a legal action following on his own immoral conduct (ibid, p. 192). These equitable maxims have evolved in order to ensure that people do not profit from their illegal or unjust acts. In the words of Dworkin, equity ensures, "that no one shall be permitted to profit by his own fraud, or to take advantage of his own wrong, or to found any claim upon his own iniquity, or to acquire property by his own crime" (Dworkin 1981, p. 23). Equity thus looks to the intention of parties, and not to the mere outward form of a transaction (Bray 2006, p. 5). As stated, this differs entirely from Aristotle's views on commutative justice that ignored intentions and motives, and attempted to ensure substantive justice purely based upon the visible or material outcomes of the case.

Equity law has provided these and many more remedies to contract law for the rectification of a variety of types of unfair or exploitative contracts, including where the contract came about by folly or mistake on the part of the grievant (Treitel 1999, p. 287).[5] The doctrines of undue influence and duress, which are to be discussed below, were developed by equity courts in order to assist parties who had been duped, coerced or influenced into detrimental agreements in situations that would otherwise have been termed legal according to the common law. For

[5]The mistake can be one of law, or of fact, or as to the value of the item (Treitel 1999, p. 287).

example, where an individual entered into a contract under the excessive influence of another, equity provided the remedy of undue influence, whereby an aggrieved person could show that she was influenced against her own interests, and even where no illegal coercion or threat (duress) existed, she would be entitled to cancel the contract (Treitel 1999, p. 375). In all cases of unlawful gain by one over another, the law of equity shines the light of enquiry onto the relationship or connection between the parties prior to the contract.

Equity law thus infused and influenced the laws relating to all forms of exploitation by one party of another in transactions, as will be discussed further below. In every case the intention is to bring justice to and to assist vulnerable parties, in the process of which issues of influence and power were scrutinized as part of the relationship between the relevant parties. The flexibility of the equitable approach enables a court to protect the reasonable expectations of parties in a way that the formal law could not (Lewis and Lobban 2003, p. 310).

To summarise, the word equity in its everyday usage can reasonably be interchanged with the words fairness and justice. The law of equity provided a body of principles and remedies that supports the most intuitive and elemental aspects of what ordinary people mean by justice and fairness, and its principles encourage judges to draw upon basic notions of fairness when the formal legal system fails to produce justice. Rawls might well have had notions of equity law in the back of his mind when he proposed reflective equilibrium as a method of arriving at a fair outcome, in which the formal principles to be applied are tested and balanced against ordinary moral beliefs, or intuitive perceptions of fairness (Dworkin 1981, p. 153). According to Dworkin, the task of moral philosophy is to provide a structure of principles that supports basic moral convictions relating to certain issues, and attempts to unify them with the legal or other relevant formal precepts (ibid, p. 155). Where RECs deliberate on complex ethical and practical matters, the reflective and holistic process embodied in the law of equity, examining all the *judicanda* and procedures in particular, could provide a useful method aimed at arriving at a just decision and outcome.

6.5 The Law of Contract, and Commutative Justice

Law is intended to be just, and the purpose of law is to provide a set of binding rules or standards that govern human interactions. Dworkin describes how rules as a form of law, such as codified criminal laws against theft or speeding, form legal standards binding upon the subjects, whilst principles, policies, and equitable legal maxims are interpreted and applied in execution of these rules and laws (ibid, p. 84). Laws operate within a structure, for example a constitution, which determines how laws are made and adjudicated on, their reach, and the procedures that are associated with every aspect thereof (ibid, p. 16). Legal systems of today still reflect elemental principles that developed in ancient Rome, based upon the foundations of the *praecepta iuris* the foundations of law, which were firstly to live virtuously, secondly to injure no-one and thirdly, to give to each man his right (Bracton 1968, p. 199). Law is broadly defined into public law on the one hand, dealing with the law of institutions, of the state and of the administration of justice, and private law on the other hand, which is primarily the contract, tort (or delict) and other forms of law that prescribe conduct between citizens (Ross 1958, p. 204). Tort is the French term for wrongdoing, a civil wrong, or wrongful act, whether intentional or accidental, from which injury occurs to another, and which includes all negligence cases as well as intentional wrongs which result in harm.

Contract law governs all voluntary transactions of exchange that give rise to legal rights or obligations, and a contract is defined as, "an agreement giving rise to obligations which are enforced or recognized by law" (Treitel 1999, p. 1). An agreement to play tennis with a neighbour, or a promise to offer a lift to a friend, or a gentleman's agreement, are examples of mutual promises but not enforceable by law and therefore not legal contracts (Hutchinson and Pretorius 2010, p. 4). Every research project on the other hand rests on a foundation of contract, which exists whether written or unwritten, which includes the signed consent forms, and which governs the rights and obligations of the stakeholders. Where multiple parties are involved, the contractual relationships between them can be interpreted both from the formal documents that are signed, as well as from or supplemented by the informal, moral and/or unwritten undertakings between them.

What distinguishes contractual rights and obligations is that they do not originate from an external source of the law, but are undertaken by the contracting parties (Smith 2002, p. 2). In every consensual act by which something is offered in exchange for something else, whether buying a bus ticket, or borrowing an umbrella, or donating blood samples, the legal rights and obligations are thus based upon a foundation of contract law. The English Law of Contract evolved over centuries from early Continental, including Prussian, French and Dutch Civil Law, all of which were formerly influenced by the Roman civil law, which included the *Law of Obligations* as enunciated by the early Roman jurists (Wessels 1951, p. 17). Over the past millennia the essentials of contract law have remained remarkably constant, replicated in different legal systems across the world and regulating all modern forms of exchange. Of the original *essentialia* to a contract identified by the Roman law jurist von Savigny in his book, *System of Modern Roman Law,* the following basic elements remain relevant and underpin contract law to this day:

1. Capacity: The parties must have the legal capacity to contract.
2. Consensus: The parties must be of the same mind and intention concerning the subject matter of the contract. In other words there must be consent, termed *concursus animorum* (a meeting of the minds) regarding all the material aspects of the agreement.
3. Formalities complete: Offer and acceptance for example must be expressed and communicated to each other, in ways that are sufficiently clear, certain and final.

4. Consideration: There must be some form or valuable consideration, or *prestatio*, (present, future or conditional benefits) offered by each party.
5. Intention to contract: There must be a serious intention to create a legal bond or *animus contrahendi* (intention to contract) on the part of the parties. (ibid, p. 15)

Modern contract law adds to the above prerequisites of a contract the conditions that the subject matter of the contract should be legal, possible, in other words capable of performance, and *certain* in that it is a definite rather than vague content (Hutchinson and Pretorius 2010, p. 6). Each party to a contract is both a giver and a receiver, as each are bound by duties as well as expectations of rights (ibid, p. 8). It is not intended to explore the further reaches of and schisms within contract law, which differs in certain respects for example between Civil, English and Roman Dutch legal jurisdictions (ibid, p. 12). The core principles as set out above guide all contracts, at the heart of which are voluntary exchanges or undertakings and/or promises made in order to bring about and define the terms of an agreement. Such an agreement is not complete and is thus neither valid nor binding until all the essential terms are clear and have been agreed.

This formulation of essential or material terms becomes relevant in the realm of research when for example DNA has been provided for a specific purpose, as set out in the research protocol and informed consent documentation. Where the informed consent form does not adequately or clearly envisage future use, for example in the Tristan da Cunha case where the contract did not cover the future sale of the DNA samples for profit, or the Havasupai case where the contract did not envisage further research on the tissue samples, such future use could be challenged as being in breach of and therefore not protected nor governed by the original terms of the contract. In all exchange transactions one of the essential terms must be the consideration or *quid pro quo* offered by the receiving party, which is colloquially referred to as the, "price tag on the promise", or as "something of value" (Treitel 1999, p. 64).

This consideration constitutes the form of benefits that are promised or shared by one party, and which can be in any legal form whatsoever. Gordan emphasises the importance of this contractual element with the hyperbolic words, "consideration is to contract law as Elvis is to rock and roll: the King!" (Gordon 1990, p. 987). In Roman contract law the one party, the owner or possessor of a 'res' or a 'thing', namely the object to be exchanged, would negotiate for a reciprocal consideration, a term referred to in English contract law simply as the price, or as a "benefit to the promisor and a detriment to the promisee" (Smith 2002, p. 69). In other words, there should always be reciprocation, or value exchanged, in the form of a received benefit. This equates with Wertheimer's view that an exchange should always involve mutual benefits (Wertheimer 1996).

Contract law is generally not concerned with the adequacy of the consideration namely, whether one party has made a good bargain or not. This contrasts with Aristotle's view of justice in exchange, in which the equivalence and fairness, mathematically calculated, is at the core of the equality and thus fairness of the exchange. When all of the aforementioned elements are in place, that is when one party has consented to part with something she chose to part with, and has received what the other party undertook to impart, namely the promised consideration or benefit, the law of contract is generally satisfied.

Consent, particularly the presence and validity thereof in respect of a certain transaction, remains a central element in legal analysis of contracts. Where consent is wrongfully obtained however, from a vulnerable party for example whose bargaining position is weak, and who enters into a disadvantageous transaction, contract law in its original form was not able to come to such party's assistance. Nor did the courts normally question the fairness or justice of a transaction, such as whether a contract is particularly one-sided, or where the parties were badly matched, unless certain proven circumstances—mistake, deceit, fraud, duress—were shown to be present. The law of contract evolved in this manner in order to support a policy of certainty and enforceability of contracts, and to enhance the stability or rule certainty of commerce. *Pacta sunt servanda* is a commonly used phrase in contract law judgements meaning that contracts in their original form should be binding (Bigwood 2003, p. 1).

In the opinion of Treitel, a formal legal system could not achieve certainty in law and justice at the same time (ibid, p. 286). It is therefore sobering to reflect on the fact that the formal body of contract law was, prior to its assimilation of the principles of equity, generally not designed to deal with vulnerability, power or capacity differentials between parties. As stated by Goodin (1985) in an article entitled, *Protecting the vulnerable,*

> There is nothing wrong with contracting power per se, nor with its exercise. For to possess power is not necessarily to use it, and to use power is not necessarily to misuse (abuse) it. Vulnerability-dependency relationships are natural, inevitable, and immutable in this world, and the power that results from them is so often a desirable feature of our intimate, economic and political relationships. (ibid, p. 192)

Another important aspect concerning consent is the fact that contract law regards all legal parties as fully autonomous *legal persons* that is persons, collectives or corporations entitled to enter into contracts in their own name, which assumes that they are capable of understanding, concluding, and consenting to binding agreements. The term *freedom of contract* describes the liberal notion that individuals should be free to enter into all and any transactions, whilst the law's protection of individuals in society, for example laws designed to protect consumers, tenants, pensions, from unfair contracts, are understood and regarded as a form of justifiable paternalism (Kimel 2003, p. 118). The idea that humans have personal autonomy, namely full control over their destiny and free choice is affected by the existence or absence of such laws (ibid, p. 129). Proper informed consent where the one party may be an illiterate individual or community is notoriously difficult to achieve in exchanges involving complex issues, such as genetic research, an issue which has led to a vast body of bioethical research (Kegley 2004; Knoppers and Chadwick 2005).

Informed consent has been deemed the bastion upon which approval for the entire subsequent clinical or research exchange transaction is grounded, yet in most cases the descriptions of the intended research are inevitably and in so many cases incomplete (O'Neil 2004, p. 1134). Here debates within bioethics and contract law draw strongly upon each other. The quality of informed consent is indeed the touchstone of a fair contract, and what constitutes both adequate disclosure of all aspects of the intended research as well as adequate comprehension of the information, the scientific details, risks and benefits, intellectual property and other considerations, remains an elusive subject. Suffice it to observe that what constitutes legal consent for the purpose of the validity of a contract is a far lower standard than an ethically satisfactory standard that would guarantee adequate comprehension in the eyes of a moral philosopher.

6.6 Exploitation in Contract Law

The purpose of this focus on exploitation in contract law is to assist with the discussion that is to follow in the next chapter on undue inducement and coercion as obstacles to benefit sharing, and in order to explain the legal underpinnings of these concepts. Following a brief overview of law and the role of contract law in governing voluntary exchanges, the three forms of contractual exploitation namely unconscionable dealing, coercion, and undue inducement will be examined in turn. Tom Beauchamp (2011) believes that the law is more influential than any other field of thought in the development of doctrines such as informed consent, yet he encourages readers to think beyond the restricted limits within which the law operates (ibid, p. 518). As he correctly states, whilst the law does not go far enough to guide morally nor to determine duties to patients and subjects regarding informed consent, it nevertheless provides the outposts and benchmarks against which the meaning of such terms are tested.

This book, in examining voluntary exchanges of human DNA for genomic research, is interested in concerns relating to fairness and exploitation. As discussed above, the law of contract evolved over centuries in order to promote the utilitarian value of certainty within the dynamic variety and realm of possible encounters. The law assumes that parties are entitled to negotiate and even contest dispassionately with a view to extracting the most advantage for themselves within the outer boundaries set by the law. Adam Smith famously opined that self-interest is the driving force behind society (Smith 1776; Gauthier 1977), believed that a normal and rational person is one who seeks to appropriate as much as possible (ibid, p. 118).

In Gauthier's view, a person is a rational agent, "if and only if he acts to appropriate as much as possible [...] and seeks more" (ibid, p. 118). Others have described normal human motivation during an exchange being to obtain, "more of the outcomes that they value and that others control" (Molm et al. 2006, p. 129). This is not to deny that parties might be imbued with sound principles including of altruism and solidarity, but rather to suggest that self-interest of individuals and institutions in all forms of exchange, including biomedical research, should not be underestimated nor regarded in any manner as pejorative. How then was the law to manage and control this self-interest, and how to prevent exploitation of one party by another, in such exchanges?

The word exploitation, as discussed in chapter two, is used more freely than it is fully understood, generally in a stigmatizing way, and directed at a party who appears in some or other way to abuse or misuse another. One of the central aspirations of the law is to establish fairness in human dealings, and the conscience of the law as applied in the courts has strived to give effect to the precept of non-exploitation (Thomas 2000; Saprai 2011). Contract law as a branch of private law has evolved functionally in order to establish certainty in exchange dealings, whilst being averse to exploitation of one by another. With the assimilation and influence of the law of equity, contract law developed three defences against exploitative contracts, namely unconscionable dealing, duress, and undue influence (Bigwood 2003, p. 2). These will be dealt with below.

Despite the fact that arguments persist amongst jurists relating to the taxonomy and legal pedigree of these defences, termed *pleas in avoidance*, their meaning and application has through centuries of application and legal judgements become increasingly clear (ibid, p. 2). Some legal jurisdictions frame these pleas differently. For example, those legal systems inspired by the Roman Dutch law group frame these pleas in avoidance under the headings of misrepresentation, including mistake, duress, and undue influence (Hutchinson and Pretorius 2010). The autonomy of contracting parties to shape their obligations as they wish is thus continually tested in the law by the evolving limits on voluntariness, namely the various ways in which the voluntary consent of parties might be restricted or constrained by the actions of others (Robertson 2005, p. 180).

The task of this book is to delve into one of the realms of private law falling under Aristotle's overall arch of corrective justice, namely the Law's response to bilateral wrongdoing or breaches of duty in contract. The context of genomic research involving indigenous peoples results in two parties being envisaged; namely research representatives on the one side, and research participants, with a focus on indigenous peoples, on the other. The focus of the enquiry is some form of exploitation in this exchange, where one party wrongfully gains, loses or is harmed in the transaction. Note that wrongful actions by one party to a transaction that are criminal, such as theft, or that lead to liability in tort, for example, where negligence or a breach of duty causes damages to one party, fall outside this enquiry. Similarly, the law relating to fraud and mistake, which also falls under the broad description of *improperly obtained consensus*, and which can also give rise to an unfair contract enabling an aggrieved party to cancel the contract, resides outside the scope of this book.

The objection to benefit sharing to be discussed in the next chapter holds that research participants might be wrongly treated, in one of these three ways. These defences operate in accordance with a principle expressed as *legal neighbourhood*, which describes the mutual responsibilities of parties belonging within a particular realm of activity, and rooted in a duty to protect the vulnerable (Bigwood 2003). According to the precepts of legal neighbourhood, the greater the power of one party, the greater the risk of exploitation of the other. Mather described the principle as being founded on Aristotle's concept of *philia,* which means, "a relationship between persons, each of whom care about the other's wellbeing" (Mather 1999, p. 57). In the field of genomic research, biomedical or pharmaceutical companies, research institutions, researchers, research participants and their communities are all encompassed in such a web of legal neighbourhood and play roles in research transactions.

6.6.1 Unconscionable Dealing

Unconscionable dealing, also called unconscionable bargain, means the unfair use of power in a contract. The doctrine emanates directly from the law of equity, and is the first of the three forms of exploitation that can occur within otherwise legal contracts. Based originally upon a notion of equitable or constructive fraud, meaning the wrongful dealing is likened to a form of fraud, the doctrine is aimed at protecting the weak from the strong, and in particular where one party is in a position to exploit a specific state of vulnerability in the other (Treitel 1999, p. 381). The doctrine is epitomised by the words of Lord Hardwicke who stated that equity would intervene in any case, "to prevent taking surreptitious advantage of the weakness or necessity of another; which knowingly to do is equally against the conscience as to take advantage of his ignorance" (Earl of Chesterfield 1751, p. 82).

Whilst no single settled formulation of unconscionableness has received general acceptance across all common law jurisdictions (Benson 2001, p. 185), it is clear that the doctrine focuses upon the dealing, meaning the manner in which the transaction was procured, rather than on the substantive results of the bargain (Bridgewater 1998).[6] Substantive unfairness of the outcome is thus not the decisive issue. Treitel (1999) describes it succinctly as a form of dealing where one party is in a position to, and exploits a particular weakness of the other (ibid, p. 386). Originating from a protective equitable jurisdiction concerned with safeguarding individuals from their own ineptness, misfortune or incompetence, it evolved towards the detection and prevention of victimisation of any form (Bigwood 2003, p. 234). Nowadays, it is used more to protect those under a disadvantage from stronger parties, who take advantage of that fact and abuse their bargaining power (ibid, p. 234).

Inequality of bargaining power has resulted in a range of different kinds of cases all deserving equitable relief (Treitel 1999, p. 383). In the case of Lloyds Bank Ltd v Bundy (1975, p. 339), the following words by Lord Denning showed the desire of equity to assist the weak, as well as the overlap between unconscionable dealing, duress and undue influence. He stated that,

[6]See list of Legal cases in References.

> English law gives relief to one who ... enters into a contract upon terms which are very unfair [...] when his bargaining power is grievously impaired by reasons of his own needs or desires, or by his own ignorance or infirmity, coupled with undue influence or pressures brought to bear on him by the other. (ibid, p. 339)

As an example, in the Canadian case of Harry v Kreutziger (1978), the owner of a boat was allowed to rescind a contract for the sale of his boat and fishing license for a nominal amount. Unknown to the seller, his fishing license was worth a great deal of money, and could have been mortgaged to finance a new boat. The court ruled that the buyer had taken advantage of the seller's lack of knowledge of the value of the license, and refused to allow the contract to be enforced. In deciding this case, in addition to examining the communication between the parties, the courts enquired into the range of acceptable practices and prices in the industry, in order to establish community standards, and whether true advantage was taken (ibid, p. 241).

In order to formulate a case based upon unconscionable dealing, the complaining party would according to the famous case of Commercial Bank of Australia v Amadio (1983, p. 463), have to show that unconscientious advantage was taken by the other party based upon the former's disabling condition or circumstances.[7] Stated more emphatically by Lord Dawson in the same case, relief is dependent upon, "exploitation by one party of another's position of disadvantage in such a manner that the former cannot in good conscience insist on the benefit of the bargain" (ibid, p. 489). The Amadio case confirmed that an aggrieved party claiming unconscionable dealing is required to show that she was (a) by some reason or circumstance placed at a special disadvantage *vis a vis* another, and (b) that unfair or unconscientious advantage was taken of the opportunity thereby created (ibid, p. 462). It was expressly stated by Judge Mason that the relief is applicable, "where an innocent party's will is overborne so that it is not independent and voluntary" and he "is unable to make a worthwhile judgment as to what is in his best interests" (ibid, p. 462).

Once a charge of unconscionable dealing is made against the allegedly stronger party, the onus is immediately placed upon the person who receives the benefit of the transaction to show that the transaction is fair (ibid, p. 474). The conditions or circumstances which might make one party more vulnerable are many and varied, however in the case of Blomley v Ryan (1956, p. 405), Judge Fullagar proceeded

[7]The Amadios guaranteed their son's indebtedness to the Commercial Bank of Australia, but when the Bank sought to enforce the guarantee, the Amadios claimed that it was unenforceable because it was unconscionable. They claimed to be at a "special disadvantage" as an equitable doctrine in Equity Law. The court ruled in Amadios' favour, taking into account the facts that the Amadios had a limited understanding of English, were not offered independent advice, and the bank knew of the Amadios' situation at the time the guarantee was signed (ibid, p. 462). The court held firstly that when the mortgage was executed the bank was aware of the Amadios' son's financial situation and knew the Amadios were not so appraised, and secondly that the bank did not advise the Amadios that there was no limit on their liability under the guarantee—the Amadios believed the liability was limited to $50,000.

to list a few: "poverty or need of any kind, sickness, age, sex, infirmity of body or mind, drunkenness, illiteracy or lack of education, lack of assistance or explanation where assistance or explanation is necessary" (ibid, p. 392). In essence, it applies where the vulnerable party is unable to judge properly for himself. The cases also protect groups such as, "the elderly and debilitated, the muddle headed, the emotionally infatuated, the tense and nervous, and the exceptionally gullible" (Bigwood 2003, p. 241), and the doctrine, triggered by an asymmetry in power advantageously used by the one, is clearly applicable to indigenous, rural or economically challenged communities from the developing world.

This form of protection is remarkably wide. It is not surprising that concerns about its scope are frequently mounted by those requiring certainty in contracts, who note amongst their complaints that judges often differ on whether a party is under a special disadvantage. For example, in both the cases of Bridgewater vs Leahy (1998) and Diprose vs Louth (1990) the judges differed from one another throughout on the question of special disadvantage. According to Bigwood, whether a party suffered from a special disadvantage is one of those concepts, such as good faith and foreseeability, that judges will continue to debate and differ on, as such concepts require careful interpretation based in each case upon known facts (ibid, p. 246).

In summary, intention and state of mind remain central to the proof of wrongdoing. It is a crucial determinant of unconscionable dealing that the stronger party knowingly and intentionally proceeds to secure advantage against the weaker party. Regarding whether the stronger party could or should have known of the other party's weakness, for example, judges in the case of Baden (1992) developed five possible levels of cognitive awareness applicable to the stronger party. These levels were described as first, actual knowledge of such a weakness, secondly wilful shutting one's eyes to the obvious, thirdly wilfully and recklessly failing to make such enquiries as an honest and reasonable person would make, fourthly knowledge of circumstances that would indicate the facts to an honest and reasonable person, and fifthly knowledge of circumstances that would put an honest and reasonable person on enquiry (ibid, p. 61).

In addition to the intention of the stronger party, the cases of unconscionable dealing all have another thing in common, namely that they need to show the advantage taking complained of to be unconscionable.[8] This word means, "serious, and arousing a sense of shock, reprehensibility or unreasonableness, so as to be sufficiently divergent from community standards of morality" (Bigwood 2003, p. 248). In other words it cannot be used by the weak to simply evade all responsibility for their dealings. As a final note on the applicability of this equitable doctrine in bringing about justice in exchange, it should be noted that the primary focus on the procedural fairness evident in the dealings of the parties is balanced with and influenced by an analysis of the substantive unfairness in the case (ibid, p. 271).

[8]Unconscionable is also described as 'not morally right' and 'excessive'. [Online] Available at http://www.merriam-webster.com/dictionary/unconscionable.

6.6.2 Duress (Coercion)

The second of the three forms of exploitation that can take place within legal contracts is called duress, also referred to as wrongful coercion. It is clearly wrong for one party to force another to enter a contract by means of an improper threat or pressure, and the law recognises that consent derived under such circumstances is inauthentic. A contract is therefore voidable if it was made under duress, for coercion of the will vitiates consent (Treitel 1999, p. 375). When an agreement is made as a result of improper pressure that amounts to intimidation, there is no lack of consensus, because the intimidated party decides, under duress, to consent. The problem however is that the consensus or agreement was improperly obtained by the making of some form of unfair or illegal threat. Under these circumstances the agreement thus obtained is deemed initially valid, based upon both parties consent, but it is voidable by the innocent or aggrieved party (Hutchinson and Pretorius 2010, p. 136).

As an example, a party might be coerced by the threat "If you do not sign this agreement, I will ensure that harm will come to your family". The threatened party in such a case can claim to have acted under duress, as a result of improper coercion, and apply for cancellation of the contract. Duress involves coercion of the mind and the will of the victim, termed *vis compulsiva* in Latin, meaning a verbal or non-physical threat. This is distinct from *vis absoluta*, which means overwhelming physical force such as when one twists a person's arm or forces her hand to sign a contract (ibid, p. 138).

The doctrine of duress is activated by the improper threat of harm, rather than the harm itself, the threat of which constitutes the element of coercion. To qualify as coercive a threat needs to be directed at the life, bodily integrity or property of the person concerned, or her immediate family (De Wet and van Wyk 1992, p. 50), the fear should be reasonable, and the harm threatened should be imminent (Wessels 1951, p. 1167). Finally, the threat of harm should be unlawful, for duress is not based upon the presence of fear per se but constitutes the obtaining of consent by improper means (De Wet and van Wyk 1992, p. 49). The issue of coercion in relation to the voluntariness of consent, from perspectives of law, ethics and morality, is the subject of a vast body of literature (Hughes 2001; Von Savigny 2012 and Largent et al. 2013), much of which lies outside the scope of this book.

Freedom of contract is not prevented or curbed by the doctrine of duress, which does not prohibit the use of pressure in human transactions. In fact it is accepted that some form of pressure, inducement, or even forceful persuasion is an endemic feature of contract bargaining activity. In an article entitled, *Bargaining, duress and economic activity,* Hale (1943) noted that all contracting involves a measure of coercion, but not necessarily duress. The former is termed the rough and tumble of hard bargaining, whereas the latter term involves pressure constituted by a threat of harm that is in some way unfair or improper. An offer of payment or an inducement to a research participant for example that contains no actual improper threat cannot be termed duress (Grady 2005, p. 1683). The classic case of duress, described by Feinberg as the "model coercer" is constituted by the person with the gun who creates an exploitable situation by using the weapon to back up the threat implied in 'your money or your life?' (Feinberg 1983, p. 208). The recipient certainly consents to hand over the money, but the consent is forced, thus not authentic. Some, for obvious reasons, also describe this form of coercion as active exploitation (ibid, p. 208).

However, where the stronger party does not in fact create the exploitable situation, what is termed *lawful act duress* comes into play. This is where the threat is to carry out an act which is clearly legal, such as a threat to report a person to the authorities for driving a motor car without a licence, or a threat to publish truthful information that will discredit, expose or embarrass the other. This threat might indeed be forceful and even unpleasant, but is in no way unlawful or improper. This lawful act duress is thus not illegal per se under the doctrine of duress, but the innate and moral wrongness of such conduct still constitutes an exploitative contract protected by unconscionable dealing; namely the previous defence against exploitation where a vulnerable person's particular weakness or situation is specifically targeted (Bigwood 2003, p. 23).

The law relating to duress has developed through legal scrutiny of countless examples of different forms of coercive pressure. In all these cases involving varying degrees of improperness on the part of the coercer, as in all the cases influenced by the law of equity, the enquiries turned on the state of mind or bilateral intentions of the parties. For example, Grantham and Rickett (2000) argue that, "the law intervenes in a case of duress, not solely because of the defendant's conduct, but because of the effect of that conduct on the plaintiff's ability to exercise a free and independent choice" (ibid, p. 185). However, they conclude that any involuntariness in the consent of the aggrieved party can only be judged relative to the wrongfulness of the conduct of the party accused of coercion (ibid, p. 187).

These conclusions are further reminders of the correlativity of exchange transactions, and the bilateral nature of justice in exchange as envisaged by Aristotle. The defectiveness of the resulting consent, or otherwise stated the impaired volition of the coerced party caused by the wrongful pressure, remains the principal justification for law's interference with an agreement between the two parties (Bigwood 2003, p. 286). Normal commercial pressure that constrains one's choices is thus legitimate, whilst it is the nature and type of pressure or coercion that the doctrine of duress regards as illegitimate. This is aptly summed up by the words of DM Campbell in Peanut Marketing Board v Cuda (1984), "duress is not a matter of a person being left no choice—it is a matter of the pressure applied being of a kind which the law does not regard as legitimate" (ibid, p. 378).

This book is concerned with the question, when does an offer that is highly persuasive and applies pressure on a clearly weaker party, such as a research proposal that contains a certain degree of risk but carries attractive benefits, amount to a wrongfully coercive proposal? When is such pressure illegitimate due to coercion? The two pronged theory of duress proposed by Wertheimer states that such pressure is only unlawful if the proposal places the weaker party in such a position that she has no reasonable alternative (Wertheimer 1987, p. 172), and most importantly if the proposal itself is unfair, illegal or otherwise wrongful, "A coerces B if and only if A's proposal creates a choice situation for B such that B has no reasonable alternative but to do X and it is wrong for A to make such a proposal to B" (ibid, p. 172).

Where a research proposal is not wrongful, and is duly approved by the relevant regulatory bodies, an offer made within the approved parameters of the proposal to elicit participation cannot be termed duress, for the proposal is lawful. If the effect of the lawful offer is however coercive and manipulative in effect, it might well be judged as lawful act duress and an exploitative contract under the unconscionable dealing doctrine. Threats and pressure of all forms, whether express or implied, legitimate or otherwise, thus still have the ability to compel compliance. Borderline examples of coercive language that is not obviously illegal might be threats such as—"you had better watch your back!"—or in the context of genetic research—"if your community does not agree to participate in this research project I predict that you will regret it". These words might on their face appear innocuous, or at the least not illegal, but the context of who is making the threat to whom is at the crux of the bipolar relationship which comes under the scrutiny of commutative justice. Such words issued from a person of known influence would form relevant factors in establishing the exploitativeness, under the doctrine of duress, and thus the illegality of the resultant contract.

Commutative justice cases involving *inter alia* duress generally entail an examination of the notional *baseline* of parties, seeking to correct the fact that one party has gained more than she should have in the exchange (Weinrib 2002, p. 355). The baseline is intended to be a description of a person's situation in life, which she should attempt to enhance or improve. Attempts to set what is termed a baseline for the person receiving the threat or offer, in order to establish whether she became better or worse off after acting on the basis of the coercion, have included a variety of standards. This includes what Wertheimer describes as, "moralised or normative, subjective, non-moral/empirical, objective, rights based", standards, including attempts to establish what is referred to as a "statistical baseline" (Wertheimer 1987, p. 207).

According to Weinrib Aristotle's account for the parties' baseline was their initial notional equality, which was his way of referring to the entitlement of each of the parties to have what was rightfully theirs rather than actual equality (Weinrib 2002, p. 354). Whilst no single test has emerged to establish clearly such a notional baseline for parties in such exchanges, courts have not been embarrassed to acknowledge that often the best they can do is to focus on the distinctive facts and features of the case in hand in order to arrive at a just conclusion (Bigwood 2003, p. 301).

The law sustains a liberal conception of contract—the freedom to contract—and coercion takes place in many forms. Defective consent on the part of a vulnerable party due to strong persuasion is on its own not sufficient to excuse that party from the contract. The application of commutative justice—a form of corrective justice—through the doctrine of duress requires the unhappy party to show how the coercion of the other party was wrongful and brought about her defective consent (ibid, p. 288). The modern doctrine of duress focuses on both the coercive conduct of the first party, as well as on the conduct's effect on the contractual assent of the second, acknowledging their respective roles in the bilateral transaction (Grantham and Rickett 2000, p. 187).

Wertheimer takes this further, after an analysis of US contract law, and suggests that such duress has less to do with freedom and voluntariness, and more to do with wrongfulness and unfairness (Wertheimer 1987, p. 53). He would therefore imply that it is the eventual effect on the weaker party's baseline rather than the coercion itself that is the ultimate test of wrongful duress. One could categorise Wertheimer's emphasis on the baseline test as focussing on the substance of the outcome, rather than on the procedure that led to the outcome. In other words he would conclude that if the end result is beneficial for the weaker party, there should be no concern about employment of coercion or other wrongful conduct. When the doctrine of duress is invoked in order to correct the injustice that exists within a voluntary exchange, it is however not necessary to decide between those two processes.

As Pogge's framework shows, material outcomes and procedural facts are in any event inextricably linked in justice. Following the precepts of equity, the procedural facts of any particular transaction in which coercive and wrongful pressure by one party results in assent which is less than voluntary and free, that is defective, should first be examined in their entirety. If the coercion is wrongful or illegal, the doctrine of duress comes to the assistance of the aggrieved party. Where the coercion is not wrongful in the form of being illegal, but it is nevertheless exploitative in essence, the victim can nevertheless be protected through the doctrine of unconscionable dealing described above.

6.6.3 Undue Influence

Undue influence as the last of the three doctrines needs to be examined, in order to understand when it is a problem or concern in genomic research. The expression, "the shepherd must not become a wolf" used by Judge Robert Megarry in Tito v Waddell (1977, p. 241), summarises the essence of this third form of legal protection against an unfair contract, which puts the trust relationship between two parties, the shepherd and the sheep, at the centre of the enquiry. Here the wrongful form of power exercised by the one who brings about the contract is not the manipulative or coercive power discussed in the previous two sections, but what is termed inducive pressure or power (Wertheimer 1996, p. 134).

As stated above, parties are entitled and even expected to exert persuasion on each other in the process of bargaining, and the precise limits of such lawful persuasion are, as in the coercion test, often difficult to determine. Judges have in undue influence cases continually resorted to what Bigwood (2003) terms, "pseudo-psychological terminology" similar to that used in coercion cases, using words such as "dominated, controlled, subordinated, subverted or overborne" to describe the effect of such power on the induced wills of the weaker party (ibid, p. 374). In cases of both undue influence and coercion, there is a superior party who wrongfully persuades the weaker party to act in a certain way.

What distinguishes undue influence is however the absence of coercive persuasion and the exploitative use or breach of a special trust or relationship. Alternatively put, there is an exploitation of misplaced trust or reliance where self-denial would be both expected and required by the influential party (ibid, p. 375). Issues of power are interchangeable with influence, and as an abstract concept, influence can be described interchangeably with the notion of power. One person exerts power over the other whenever the former coerces, misleads, exploits or manipulates and thereby influences that other (ibid, p. 180). Influence itself is not frowned upon in the law, as stated by Kekewich, J. in the case of Allcard v Skinner (1887),

> Nay, it recognises influence as natural and right. Few, if any, men are gifted with characters enabling them to act, or even think, with complete independence of others ... But the law requires that influence, however natural and right, shall not be unduly exercised. (ibid, p. 193)

Whereas in coercion, the will of the victim is altered and affected by the coercive threat, in true cases of undue influence, the victim believes she is acting in her own best interests. Rather than being motivated by fear, she believes that the influencing party is acting in her best interests, rather than his own (Fingarette 1985, p. 105). The victim of undue influence is to an extent naive or innocent, and her consent is both real and voluntary, albeit wrongly induced. The unduly influenced party thus enters into the agreement apparently voluntarily, rather than feeling under threat, in protest or even aware of being wrongfully influenced. For that reason, and following Aristotle's insistence on the relational core of corrective justice, the investigation of the court is focussed upon the particulars of the relationship between the parties, and the opportunities that the relationship might provide for abuse, rather than on the material aspects of the transaction. As stated by Judge Dixon in the early case of Johnson v Buttress (1936, p. 134), the foundation of the setting aside of such a transaction lies in, "the prevention of an unconscientious use of any special capacity or opportunity that may exist or arise of influencing (P)'s will or freedom of judgement in reference to such a matter" (ibid, p. 134).

As stated earlier, the origin of undue influence lies in the English law of equity. The more narrow doctrine of duress only assisted victims in cases of wrongful coercion, and exposed the need for equitable relief where no overt and wrongful pressure was involved. The courts of equity consequently developed the more elastic doctrine of undue influence (Hutchinson and Pretorius 2010, p. 141), the nature and rationale of which is based squarely upon the fiduciary relationship, the elemental quality of the sheep and the shepherd image referred to above. In the early and famous case of Allcard v Skinner (1887, pp. 157 and 168), which provided many of the key principles underlying the doctrine, Judge Kekewich stated,

> The law does not prohibit persons making gifts or conferring benefits upon other persons standing in a position of trust or confidence towards them, whether from affection, gratitude or otherwise, as long as the gifts or benefits are the voluntary and well understood acts of such persons. (ibid, p. 157)

This focus on the right to donate, altruistically or otherwise, as well as on the fiduciary nature of the relationship, are both important aspects of the issues being explored in this book, namely the provision of human tissue for genetic research. What is at the core of justice in all these transactions is the necessity of such exchange to be, in the words of Judge Kekewich, fully "voluntary and well understood" by the weaker party. In the same case, Judge Bowen added the following words, indicating the true focus of the entire enquiry being on the elusive realm of the conscience of the more powerful or inducing party and his or her potential to influence others, "the undue influence jurisdiction is not a limitation placed upon the action of (P): it is rather a fetter placed upon the conscience of (D) and one which arises out of public policy and fair play" (ibid, p. 190).

This confirms the focus of the enquiry on the influential power of the stronger or trusted party, and the use of that power to the detriment of the influenced, or trusting party. In undue influence cases, and reflective of the aspirations of the law of equity to seek fairness in its reach into private transactions between consenting adults, the word undue is freely interchanged with similarly pejorative adjectives such as wrongful, abusive, unfair or improper, used to describe the exercise of influence. This again requires the enquiry to go beyond the mere outward or material form of the contract and the facts of a case, and to explore the complex internal arena of morals, motives and conscience. In supporting an analysis of all relevant facts, Lord Clyde suggested that, "undue influence is something which can be more easily recognised when found than exhaustively analysed in the abstract," (Royal Bank of Scotland plc v Etridge 2001, p. 1050). Of the many attempts to provide a workable test for undue influence, it was concluded in the case of Patel v Grobelaar (1974, p. 532), that a party that seeks to set aside a contract on the grounds of undue influence must establish three facts, namely:

1. That the other party obtained an influence over him or her
2. That this influence weakened his or her powers of resistance and rendered his or her will compliant
3. That the other party used this influence in an unscrupulous manner to persuade him or her to agree to a transaction that (a) was prejudicial to him or her and (b) which he or she would not have concluded with normal freedom of will.

In the case of Allcard v Skinner (1887), the development of the law of undue inducement into two distinct heads or classes was classically expounded by Lord Lindley. In the first class, namely actual undue influence, the opportunity to exert influence is not located in the status of the relationship per se, but is located in a particular situation brought about by the wrongful party. In this case Miss Allcard joined a religious order that required poverty, chastity and obedience, and was introduced to Miss Skinner (a lady superior of the religious order) who encouraged her to donate all of her property in her will into Miss Skinner's name. One of the rules of the order was that the voice of the superior is the voice of God.

When Miss Allcard left the religious order, she claimed return of her property, on the basis that the lady superior had unduly influenced her. The court was satisfied on the actual facts, rather than on the basis of a presumption based on the formal relationship between the parties, that the gift was the result of influence, "expressly used by the done for the purpose" (ibid, p. 181). Miss Skinner denied having expressly influenced Miss Allcard, but the judges held that the facts showed that she did. This form of undue influence is very closely linked to the other two, but the key distinguishing factor is the actual reposing by one party of confidence in the other (Treitel 1999, p. 379), where "some overreaching, some form of cheating, and generally though not always, some personal advantage obtained" (Allcard v Skinner 1887, p. 181). Lord Lindley also added, "insidious forms of spiritual tyranny" to his descriptions of conduct to be guarded against (ibid, p. 182).

The second form or class of undue influence expounded by Lord Lindley, termed presumed, or relational undue influence, is the one perhaps more strongly associated with the legal doctrine's classic origins. Relationships were a particular concern in the Chancery courts of equity, and justice was brought to rectify unsatisfactory transactions particularly where relationships bore some relation to the shepherd and sheep image described above. As was set out by Lord MacNaghten in the Bank of Montreal v Stuart case (1911, p. 137), where such a relationship existed, the law would presume that any questioned gift of immoderate or irrational size from the victim to the person in the position of trust or power was wrongful, and the burden of evidence to remove the presumption would immediately shift to the influencing party. Such relationships that would lead to a presumption of undue influence in any disputed transfers include parent and child, clergy member and communicant, physician and patient (Farnsworth 1999, para 4.2). These groups have also included teachers, doctors, lawyers, bankers, trustees, priests, investment advisors, spiritual leaders and more (Treitel 2003, p. 409).

As an example of this form, in a case where a client purchased an asset from her solicitor, the onus was on the solicitor to show that the client was fully informed of all the relevant facts, that the client was separately advised, and that the transaction was a fair one (Wright v Carter 1903). Sometimes termed *status based fiduciaries*, such a presumed relationship of trust is as a consequence of a formal role and function which results in them being automatically trusted, such as trustees vis a vis their beneficiaries, physicians and their patients (Finn 1989, p. 33). Shepherd describes this sort of fiduciary relationship as existing, "whenever any person acquires a power of any type on condition that he also receive with it a duty to utilise that power in the best interest of another" (Shepherd 1981, p. 96). The law thus presumes, from the combination of the fact of the relationship and of the transaction, that the opportunity for unfair advantage taking was in fact real. Lord Herschell set out the basis of this assumption in the case of Bray v Ford (1986, p. 52), where he stated that fiduciaries could be devious, furtive and shrewd.

It is also important to note that it is no excuse for the person charged with undue influence to state that the dealing was in the victim's interests, or that she being the accused party had not received benefit from it (Treitel 1999, p. 377). Such well-meaning paternalism is of no benefit to the party accused of undue influence. The fact that the victim's will was subverted, even unknowingly, means that even substantively fair commercial transactions entered into as a result of relational undue influence should be set aside, not only as a matter of right but additionally as a matter of justice (Allcard v Skinner 1887, p. 183). In summary, substantive unfairness in the form of a bad outcome is not integral to a case of relational undue influence, and it is no defence in such cases that the disputed transaction might have turned out well for the victim. As Deane J stated,

> In such cases, we do not ask whether it was a good bargain or a bad bargain before we set it aside. The mere fact that you, being in circumstances which made it your duty to give your client advice, have put yourself in the position [...] where your interests come in conflict, that mere fact authorises him to set aside the contract if he chooses to do so. (McPherson v Watt 1877, p. 272)

Undue influence should thus remain an influential doctrine to assist vulnerable parties, being applicable even when the exploitation is subtle and not manifestly obvious. It is clear therefore that within the legal neighbourhood comprising stakeholders within the biomedical research community, all of the factors that give rise to caution are present. Research institutions interact with communities in transactions that are often less than straightforward, and the communities are vulnerable in that they defer to and rely entirely upon the former party for crucial information. Researchers are educated people, often with a high status relative to the community. The notion of vulnerability in cases of relational undue influence is thus highly relevant, as with other forms of exploitative contract, but it is more importantly linked specifically to the relationship of trust, which a participant community can quickly acquire with a researcher, especially if they believe that the research is medical and might be linked to the provision of health care.

The culpability of the trusted party, known in legal terminology as the fiduciary, is that she has knowledge of the beneficiary's particular susceptibility towards her influence, which she proceeds to exploit. To trust a person is to render oneself vulnerable to that person, who acquires a certain corresponding power over the trusting one. The greater the trust, the greater the interpersonal vulnerability to the other's power. What Bigwood (2003) terms equity's "prophylactic stance against possible wrongdoing by a fiduciary" (ibid, p. 404), is thus triggered by the inherent inequality and potential for abuse inherent in all status-based relationships of trust. The consent provided by the vulnerable party where such a relationship of influence exists, is thus closely examined by the law's procedural enquiry for signs of deference of voluntariness.

Concluding the discussion on undue influence, it is, together with unconscionable dealing and duress; inimical to the existence of the free agency required to constitute legal voluntariness and a person's ability to freely assent. The precise bounds of undue influence overlap with the other exculpatory doctrines of duress and unconscionable dealing discussed above, in addition to the doctrines of mistake, breach of fiduciary duty and misrepresentation (Bigwood 2003, p. 378). What is required for the doctrine of undue influence to assist an aggrieved party is firstly, a relationship of influence, and secondly a suspicious transaction (ibid, p. 424), the latter meaning a transaction that is not clearly in the interests of the aggrieved party. In all cases the weaker party is justified in trusting that the other

Table 6.1 Three doctrines of contractual exploitation

Three contractual forms of exploitative dealing	The role and nature of consent	The relationship between the parties	The essence of this form of exploitation	The particular form of vulnerability
Unconscionable dealing	Real consent exists, but its voluntariness is impaired by the dealings of the other party. (Motivated by need or circumstances)	No former relationship required between the parties	Taking unfair advantage of a vulnerable party, to achieve her real consent, for the benefit of the more powerful party; i.e. dealing in a wrongful manner	Circumstances not of the stronger party's making result in a vulnerability of the weaker party
Duress	Real consent exists, but its voluntariness is impaired as it takes place (is entirely forced) as a result of the illegal coercion by the other party. (Motivated by fear)	No former relationship required between the parties	Improper use of coercion or illegal pressure in order to bring about consent, to the benefit of the coercer	The improper/illegal use of coercion or pressure by one party makes the other party vulnerable and secures forced consent
Undue influence	Real consent exists. However its voluntariness is impaired by the undue influence of the other. (Motivated by trust and influence)	The relationship of trust or influence between the parties is central to the presence of undue influence	Improper use of trust relationship in order to achieve the real consent of the influenced person, in a 'suspicious transaction'	The relationship of trust makes the trusting party particularly vulnerable in respect of influence of the trusted party

will not act contrary to the former's best interests (ibid, p. 428). Biomedical research clearly provides situations where communities are specially disadvantaged in relation to a researcher (Flanagan 2000, p. 314), in that they place reliance on information provided by that trusted person. Undue influence is a powerful and flexible tool for justice in such cases.

The three forms of contractual exploitation are distinguished from one another on the basis of characteristics determined in a holistic procedural assessment. The following diagram reflects the most important characteristics of the three doctrines (Table 6.1).

6.7 Conclusion

In concluding this chapter on justice and the role of the law in its achievement, and prior to examining the warnings against undue influence and coercion contained in the bioethics guidelines, the briefest recapitulation of the above discussion is appropriate. What was required was an understanding of the concepts of undue influence and coercion, within a broad framework of justice and law. A glance over the realm of international justice and the place of law within the quest for fairness and equity led to Pogge's conceptual depiction of justice, which was followed by an examination of the early foundations for law, justice and equity as laid down by Aristotle, where he divided forms of justice primarily into distributive and corrective justice. Corrective justice which concerns bilateral actions between humans, includes commutative justice, otherwise known as justice in exchange, which frames the discussions and concerns raised in this book. Mayer's second form of exploitation, it should be recalled, a non-legal conceptualisation of exploitation, is similarly situated within such voluntary and bilateral exchanges.

Modern law was then briefly reviewed as the manifestation of the attempt by states to both regulate and bring justice to dealings between people, commencing with an analysis of equity law followed by contract law, which regulates all voluntary exchange transactions. Finally, the three areas of contract law that were most strongly influenced by the law of equity and that were developed specifically to remedy exploitation of vulnerable parties were analysed in some detail. Each of these forms of contractual exploitation takes place within the context of the 'legal neighbourhood' of genomic research, which includes amongst all the other stakeholders, researchers and indigenous communities. This legal neighbourhood it will be recalled requires justice in exchange in all cases where a certain vulnerability exists on the one side of a transaction, where that vulnerability or dependency is known to the other, who then uses that power in the dealing (Bigwood 2003, p. 506). This book suggests that biomedical companies, genetic bio-banks, researchers, research participants and their communities are all encompassed within such relationships of legal neighbourhood, and when transacting in the processes of genetic research are bound by duties of equity and fairness towards one another. Bearing in mind the core components of and distinctions between the types of contractual exploitation as depicted in the figure above, the discussion will now turn to address whether and under what circumstances benefit-sharing might either coerce or unduly influence potential participants in genomic research.

References

Allcard v Skinner (1887). AC 36 Ch D 145 [Legal Case].

Aristotle 322 (1985). *Nicomachean ethics* (Irwin. T, Trans.) London: Hackett Publishing.

Aristotle (1094a). *Nicomachean Ethics.* (H. Rackham, Trans.) Cambridge, Massachusetts: Harvard University Press.

Baden v Societe Generale pour Favoriser le Developpement du Commerce et l'Industrie en France (1982,1992) All ER 161.

Bank of Montreal v Stuart (1911) AC 120.

Beauchamp, T. L. (2011). Informed consent: Its history, meaning and present challenges. *Cambridge Quarterly of Healthcare Ethics, 20*(4), 515–523.

Benson, P. (2001). The Unity of Contract Law. In P. Benson (Ed.), *The theory of contract: New essays*. Cambridge: Cambridge University Press.

Bigwood, R. (2003). *Exploitative contracts.* Oxford: Oxford University Press.

Binmore, K. (2006). *Origins of fair play: Volume 614 of papers on economics and evolution.* Max Planck Institute of Economics. https://papers.econ.mpg.de/evo/discussionpapers/2006-14.pdf.

Blomley v Ryan (1956). 99 CLR 362 HCA 81.

Boylan, M. (2011). *Morality and global justice: Justifications and applications.* Boulder: Westview Press.

Bracton, Henry de. (1968). *On the laws and customs of england.* (S. Thorne, Trans.). Cambridge: Belknap Press of Harvard University Press.

Bray, Judith. (2006). *Equity and trusts.* London: Hodder Arnold.

Bray v Ford (1986) AC 44.

Bridgewater v Leahy (1998) 194 CLR 457.

Brownlie, L. (2003). *Principles of public international law.* New York: Oxford University Press.

Chun, S. (2010). Natural justice and its political implications: Legal philosophy revealed in the doctrine of the mean. Culture Mandala: The Bulletin of the Centre for East-West Cultural and Economic Studies. Vol. 9, (1) Article 55.

Commercial Bank of Australia Limited v Amadio (1983) 151 CLR 447.

Cooney, W., (1998). Rights theory. In *Encyclopedia of Applied Ethics* (Vol. 3). San Diego: Academic Press.

Deutsch, M. (1985). *Distributive Justice.* New Haven: Yale University Press.

De Wet, J.C., van Wyk, A.G. (1992). *Die Suid Afrikaanse Kontraktereg en Handelsreg* (5th ed., Vol. 1.). Berea: Butterworths.

Diprose v Louth (No 2) (1990) 54 SASR 450.

Dugard, J. (1994). *International law: A south african perspective.* Juta and Co: Cape Town.

Dworkin, R. (1981). *Taking rights seriously* (3rd ed.). London: Gerald Duckworth and Co.

Earl of Chesterfield v Janssen. (1751). 2 Ves Sen 125:28 ER 82.

Farnsworth, E. A. (1999). *Contracts* (3rd ed.). London: Aspen Publishers.

Faunce, TA., (2005). Global intellectual property protection of innovative pharmaceuticals: Challenges for bioethics and health law. In B. Bennett & G.F. Tomossy (Eds.), *Globalization and health: Challenges for health law and bioethics*, Chap. 7 (p. 87). Springer: Netherlands.

Feinberg, J. (1983). Noncoercive exploitation. In R. Sartorius (Ed.), *Paternalism* (pp. 201–235). University of Minnesota Press: Minneapolis.

Fingarette, H. (1985). Victimisation: A legal analysis of coercion, deception, undue influence and excusable prison escape. *Washington and Lee Law Review, 42*, 65–118.

Finn, P. D. (1989). The Fiduciary Principle. In T. G. Youdan (Ed.), *Equity, Fiduciaries and Trusts*. Toronto: Carswell.
Flanagan, R. (2000). Fiduciary regulation of sexual exploitation. *Canadian Bar Review*, 301–314.
Folger, R. (Ed.). (1984). *The sense of injustice: social psychological perspectives*. New York: Plenum Press.
Franck, T. M. (1995). Fairness in international law and institutions. New York: Oxford University Press.
Gauthier, D. (1977). The social contract as ideology. *Philosophy & Public Affairs, 6*, 130–164.
Goodin, R. E. (1985). *Protecting the vulnerable: A reanalysis of our social responsibilities*. Chicago: University of Chicago Press.
Gordon, J. D. (1990). A Dialogue about the doctrine of consideration. *Cornell Law Review., 75*, 987–1006.
Grady, C. (2005). Payment of clinical research subjects. *Journal of Clinical Investigation, 115*, 1681–1687.
Grantham, R. B., & Rickett, C. E. E. (2000). *Enrichment and restitution in New Zealand*. Oxford: Hart.
Hale, Rl. (1943). Bargaining, duress and economic liberty. *Columbia Law Review, 43*(603), 28.
Halgarth, M. (1998). Consequentialism and deontology. In *Encyclopedia of the ethics of new technology* (Vol. 1). San Diego, CA: Academic Press.
Hanbury, H. G. (1981). *Hanbury and maudsley: Modern equity*. London: Stevens and Sons.
Harry v Kreutziger (1978) 95 DLR (3rd) 231.
Hoexter, C. (2002). *The new constitutional and administrative law*. Juta: Cape Town.
Hughes, SS. (2001). Making dollars out of DNA: the first major patent in biotechnology and the commercialization of molecular biology 1974–1980. Isisi, 541–575. http://wwwdigitaliassets.lib.berkely.edu/roho/ucb/text/makingdollarsoutofdna.
Hutchinson, D., & Pretorius, C. (2010). (Eds) Law of Contract in South Africa Cape Town: Oxford University Press.
Johnson v Butress (1936) 56 CLR 113.
Kegley, J. A. K. (2004). Challenges to informed consent. *European Molecular Biology Organisation, 5*(9), 832–836.
Kidder, R. (2003). How good people make tough choices. *Resolving the dilemmas of ethical living*. New York: Harper.
Kimel, D. (2003). *From promise to contract: towards a liberal theory of contract*. Oxford: Hart Publishing.
Knoppers, B. M., & Chadwick, R. (2005). Human genetic research: emerging trends in ethics. *Nature Reviews Genetics, 6*, 75–79.
Largent, E., Grady, C., Miller, F.G. & Wertheimer, A. (2013). Money coercion and undue inducement: Attitudes about payments to research participants. *IRB: Ethics and Human Research* 34(1)(2012), 1–8.
Lewis, A., & Lobban, M. (2003). *Law and history: Current legal issues* (Vol. 6). Oxford: Oxford University Press.
Lloyds Bank vs Bundy (1975) EWCA Civ 8.
Luban, D. (2001). Law's Blindfold. In M. Davis & A. Stark (Eds.), *Conflict of interest in the professions*. Oxford: Oxford University Press.
Mather, H. (1999). *Contract law and morality*. Westport Connecticut: Greenwood Press.
Mayer, R. (2007). What's wrong with exploitation? *Journal of Applied Philosophy,* 24(2), 137–150.
McPherson v Watt (1877) LR 3 App Cas 254.
Molm, L., Collet, J., & Schaefer, D. (2006). Conflict and fairness in social exchange. *Social Forces* 84(4), 2331–2352.
O'Neill, O. (2004). Informed consent and public health. *Philosophical Transactions of the Royal Society: Biological Sciences., 359*(1447), 1133–1136.
Patel vs Grobelaar (1974) SA 532 A.

Peanut Marketing Board v Cuda (1984) 79 FLR 368.
Pogge, T.W. (2006). Justice. In D.M. Borchert (Ed.) *Encyclopedia of philosophy*. (2nd Ed., pp. 862–870). Detroit: Macmillan.
Robertson, A. (2005). The limits of voluntariness in contract. *Melbourne University Law Review, 29*(1), 179–212.
Ross, A. (1958). *On law and justice*. London: Stevens and Sons.
Royal Bank of Scotland plc v Etridge (no 2) (2001) 3 WLR 1021.
Saprai, P. (2011). Weinrib on unjust enrichment. *Canadian Journal of Law and Jurisprudence, 24*(1), 183–192.
Shepherd, J. C. (1981). *The law of fiduciaries*. Toronto: Craswell.
Smith, A. (1776) (1976 edition). An Enquiry into the nature and causes of the wealth of nations. In R. H. Campbell & A. S. Skinner. Oxford: Clarendon Press.
Smith, J. C. (2002). The law of contract. London: Sweet & Maxwell.
Thomas, E. W. (2000). An endorsement of a more flexible law on civil remedies. *Waikato Law Review, 7*, 23–48.
Tito v Waddell (1977). No 2: Chancery 106.
Treitel, G. H. (1999). *Law of contract* (10th ed.). London: Sweet and Maxwell.
Treitel, G. H. (2003). *Law of contract* (11th ed.). London: Sweet and Maxwell.
Tyler, T. R. (1987). Procedural justice research. *Social Justice Research., 1*(1), 41–66.
UN (1945). United Nations Charger. http://www.ourdocuments.gov/doc.php?flash=true&doc=79.
Vlastos, G. (1962). Justice and Equality. In R. Brandt (Ed.), *Social justice*. Englewood Cliffs, NJ: Prentice-Hall.
Von Savigny, F. K. (2012). System of modern roman law. (W. Holloway, Trans.). Madison: the University of Wisconsin.
Wallace, R. M. M. (2005). *International law*. London: Sweet and Maxwell.
Weinrib, L. L. (1987). *Natural law and justice*. Cambridge: Harvard University Press.
Weinrib, E. J. (2002). Corrective justice in a nutshell. *University of Toronto Law Journal, 52*, 349–356.
Wertheimer, A. (1987). *Coercion*. Princeton University Press: Princeton.
Wertheimer, A. (1996). *Exploitation*. Princeton University Press: Princeton.
Wessels, J. W. (1951). The law of contract in South Africa (2nd ed.). Durban: Butterworth.
Wilke, M. (2012). The world health organizations pandemic influenza preparedness framework for the sharing of influenza viruses and access to vaccines and other benefits. ABS in the health arena. http://www.feu.unibremen.de/cms/fileadmin/dokumente/Workshop/Abstracts/Marie_Wilke_WHO_pandemic_influenza_preparedness_framework.pdf.
Winthrop, D. (1978). Aristotle and theories of justice. *American Political Science Review, 72*(4), 1201–1216.
Worthington, S. (2006). *Equity* (2nd ed.). Oxford University Press: Oxford.
Wright v Carter (1903) 1 Ch 27.

Chapter 7
Undue Inducement and Coercion

Abstract Undue inducement and coercion, both wrongful ways of eliciting consent, are introduced as the significant impediments they are to research in the developing world. A prevailing 'international orthodoxy' is described by Weinrib as being the tendency of research ethics committees to avoid benefit sharing for fear of breaching ethical guidelines that warn against this danger. The general discussion on the undue inducement in bioethics is followed by two case studies designed to emphasize its salient features, as well as the key concerns of the international research ethics guidelines, namely vulnerability, autonomy and risk (beneficence and non-maleficence). The role, function and capacity of research ethics committees to assure beneficence and non-maleficence of genomic research are then discussed in the light of concerns about protection of research participants from undue influence or coercion. Finally the two case studies are reviewed in the light of the foregoing analysis. The discussion concludes that benefit sharing should not be limited by concerns about undue influence and coercion in genomic research, which was never the intended domain for these particular ethical concerns. The guidelines provided by the law of equity for identification of these forms of exploitation are acknowledged, which if applied by functioning research ethics committees should guarantee research participants of ethical treatment.

Keywords Undue inducement · Coercion · Bioethics · Vulnerability · Benefit sharing in genetic research · Equity law · Research ethics committees

The final impediment to benefit sharing in genetic research now stands to be addressed. This objection is expressed in the warnings contained in the research ethics guidelines, namely that offered benefits or payments might either coerce or unduly influence vulnerable research participants to consent to providing their genetic resources, against their interests and without due consideration of the potentially harmful consequences. Protection of potentially vulnerable research participants from exploitation is the ethical backbone of biomedical research, from whom the obtaining of informed consent receives much attention. Guideline 6 of

R.S. Chennells, *Equitable Access to Human Biological Resources in Developing Countries*, DOI 10.1007/978-3-319-19725-8_7

the CIOMS *International Ethical Guidelines for Biomedical Research* headed "obtaining informed consent" commences with the words, "Sponsors and investigators have a duty to refrain from unjustified deception, undue influence, or intimidation" (CIOMS 2002, p. 6), and the *Declaration of Helsinki* prior to its most recent version in October 2013 described vulnerable populations as "those who may be vulnerable to coercion and undue inducement" (WMA 2008).[1]

Undue inducement and coercion, both wrongful ways of eliciting consent, are frequently cautioned against in the plethora of ethical guidelines as discussed further below.[2] In order to enable a better understanding of this objection against benefit sharing, the previous chapter provided a contextual background to the origins of these terms both in justice as well as in law. The distinction drawn by Aristotle between what he termed distributive and corrective justice enabled this book' focus on the latter category, within which commutative justice is concerned with fairness in voluntary exchanges between two parties. A brief exposition of the laws that govern voluntary exchanges was commenced with an account of the law of equity, which influenced the common law of contract, and produced three defences against exploitative contracts, namely unconscionable dealing, coercion and undue influence. The criteria that evolved in contract law for determining the presence of exploitation in an exchange will be drawn upon in the discussion below.

It is regarded as a concern by many that undue inducement and coercion have become something of an international orthodoxy (Wilkinson and Moore 1997), presenting an impediment for research in the developing world (Grant 2002). Whilst much of the earlier academic debate on coercion and undue influence is specifically directed at the situation of research participants in clinical trials, the ethical principles at stake are also broadly applicable to individuals as well as collectives recruited for genetic research, the focus of this book. An additional predicament for research conducted in the developing world however exists for the reason that, in resource poor settings, the provision of virtually any form of benefits can theoretically be and often is in practice interpreted as an undue inducement (Chokshi et al. 2005, p. 5). For instance, Annas and Grodin (1998) have argued that in the absence of health care, virtually any offer of medical assistance, even in the guise of research, will be accepted as better than nothing.

The purpose of this chapter is to replace the lack of clarity referred to by Chokshi et al., with a clear understanding of the meaning and application of undue inducement and coercion in genetic research. It is proposed to explain how the terms should be understood and applied by RECs, to provide useful guidance,

---

[1]The most recent version of the Declaration of Helsinki published in October 2013 has removed this particular definition of vulnerability, discussed further below.

[2]The most prominent international research ethics guidelines include the following: Council for International Organisations of Medical Science and the World Health Organisation, International Ethical Guidelines (CIOMS 2002, guideline 7), UNESCO Universal Declaration on Bioethics and Human Rights Article 15 (UNESCO 2005), Declaration of Helsinki (WMA 2013) and Nuffield Council on Bioethics, the Ethics of Research Related to Healthcare in Developing Countries (2002).

and in particular to clarify whether or in what circumstances these concerns should stand in the way of the provision of benefits to research subjects in genetic research. This examination of undue influence and its two associated doctrines of coercion and unconscionable dealing will commence with a brief background to the concepts, during which two case studies will be provided to serve as benchmarks for the discussion. Next, a brief survey of selected extracts from biomedical research ethics guidelines will be conducted, in order to assess what guidance is currently available for RECs and other institutions that are required to approve genetic research, and in order to highlight the key ethical concerns.

The three most significant and interrelated ethical concerns or problems reflected in the guidelines relating to undue influence and coercion of research participants will then be examined in turn, situated within the causal triangle between vulnerability, autonomy or voluntariness, and risk of harm. Firstly, the vulnerability of research participants, and secondly, their autonomy and voluntariness during research exchanges where inducements are involved, will be discussed. Thirdly, the book turns to the issue of risks in relation to genetic research, including the relevance of such risks for communities and groups targeted for genetic research, and the central role of RECs in ensuring non-maleficence and beneficence. To conclude the analysis, the case studies will then be examined in the light of the earlier treatment of undue inducement and coercion, including the key indicators provided by the law of contractual exploitation. The chapter will close with a discussion and some ventured conclusions.

7.1 Background to Undue Inducement and Coercion in Bioethics

Godfrey Tangwa (2007) explains in an article on research with vulnerable human beings that the exploitation of human vulnerabilities is manifested in coercive research, deceptive research, and inductive research. He goes on to state,

> Inducement is a serious problem in present day research, especially in the developing world. It cannot properly be fully appraised without taking into consideration the distinction between the agent/patient of moral action. A high burden of disease, combined with desperate poverty and ignorance makes people highly vulnerable to inductive research. (ibid, p. 518)

As background to this discussion on undue inducement it should be made clear that this book excludes from its purview any payments made to individuals or groups in the form of reimbursements, medical services, or compensation for inconvenience and expenses as set out in guideline 7 of CIOMS (2002) as well as HUGO's statement on benefit sharing (HUGO 2000a, b, Paragraph G). In genomic research such 'payments for donors' as compensation for inconvenience or expenses are likely to be insignificant, and are not regarded as ethically problematic. The type of benefits that this book anticipates as possible forms of inducement are the range of collective benefits described in the appendix to the Nagoya

Protocol, which include in addition to various forms of monetary payments, seventeen forms of possible non-monetary benefits[3] (CBD 2010).

When are the persuading or influencing of others coercive or otherwise undue in an exchange between two parties? It is a natural defence or excuse of a disgruntled person who regrets some transaction, such as a child accused of a bad deed, to blame the influence of another. "She made me do it," or "I only did it because I trusted her advice." The basic inference in these excuses is some form of unseemly manipulation or exploitation of one by the other's influence. Persuasion, influencing and inducements however form part of our everyday lives. From promises of rewards for children to do their homework, to salaries and desirable incentive bonuses at the workplace, to the bargaining of attractive terms in the process of negotiating and promoting economic and other exchanges.

The question arises as to when and why are the various benefits, inducements or incentives, which are offered to humans in the course of research, in breach of the ethical line. Moreover, when is an offer or overture made by one to another, in the words of Grant and Sugarman (2004), "excessive, unwarranted, inappropriate or improper?" (ibid, p. 725). In addition, in view of the perennial need for research to advance biomedical science, how indeed are research subjects to be recruited effectively, but without coercion or undue inducement and thus in breach of the guidelines? (ibid, p. 726). Let me introduce an example of the recruitment dilemma. When a group of medical professors from Harvard conceived of a novel way of getting individuals to have their blood tested for HIV in South Africa, by inviting the public to take part in a competition where prizes worth $10,000 were to be won, the South African Medical Association strongly condemned the popular campaign, finding it in breach of the ethical principle against undue inducement. The competition was duly terminated, despite the fact that the project was acknowledged as being both worthwhile in purpose and successful in outcome (Cape Argus 2011).[4]

The previous chapter discussed the three defences developed over centuries by the English equity courts to assist any individuals wrongly manipulated, coerced or induced into detrimental contracts. A party complaining that she was persuaded or encouraged into an unsatisfactory agreement as a result of the overwhelming influence of another could for example have the transaction reversed in an equity court by claiming undue influence (Treitel 1999, p. 376). Similar defences came to the aid of vulnerable parties under the other two equitable doctrines, namely unconscionable dealing and duress. The law of contract adopted these legal doctrines, and developed relatively clear criteria for determining when a transaction

[3]Non-monetary benefits under the Nagoya Protocol include (a) sharing of research and development results, (b) collaboration, cooperation in research and development programmes, (c) participation in product development, (d) education and training (e) Admittance to ex situ facilities of genetic resources and to databases, (f) transfer of knowledge and technology, (g) strengthening capacity for technology transfer (h) institutional capacity building and more (CBD 2010, annex).

[4]Notably, the principal objection expressed against the project was that the enticements, namely the possibility of winning prizes, were declared to be in contravention of ethical conduct, and more specifically that they were "undue" (ibid, p. 3).

is deemed exploitative of a weaker party. It was explained that a legally competent individual, that is not a minor or otherwise incompetent person, is generally presumed to operate under a legal presumption of voluntariness, in terms of which such an individual was bound by the terms of a contract unless she could prove the existence of duress or other manipulation (ibid, p. 286).

The more recent history of undue inducement and coercion in the field of bioethics rather than of law is closely associated with the evolution of the doctrine of informed consent from the 1950s to the present time (Beauchamp 2012, p. 515). The various abuses and atrocities that took place involving experimentation on human beings during the 1970s and 1980s, such as the Willowbrook case referred to below resulted in a crackdown and regulatory response against coercion or undue influence in research (Emanuel 2003). One can assume that such abusive research was the primary target of the protective response. Voluntariness in consent began to be scrutinised under the law relating to informed consent (Appelbaum et al. 2009, p. 11), which held that although decision makers are susceptible to multiple influences, voluntary decisions should reflect the will of the decision-maker, and consent would be inauthentic if given under duress or some such outside force (ibid, p. 11).

A curious conundrum therefore persists in biomedical research, namely that projects in less affluent settings may be prevented or discouraged from sharing benefits with their relatively poor research participants by the strict application of the undue inducement prohibition. This is another form of double standard in medical research, in which the strict imposition of ethical standards formulated in a developed world ethos has the potential to disadvantage the developing world (Macklin 2004). In the words of Chadwick et al. (2001), there is a perceived danger that emphasis on the distribution of benefits in such a research setting might be seen not as an exercise in distributive justice, but rather as an attempt to buy people off (ibid, p. 321). If applied incorrectly, or out of an excess of caution, this lack of understanding of undue inducement might cause benefits or incentives as part of developing world genetic research projects to be wrongly rejected, thus preventing the fair allocation of benefits for those humans whose genetic samples are required.

The underlying ethical concern thus resides in the supposition that providing payment or other benefits can, under certain circumstances, unduly influence, coerce or otherwise manipulate prospective research participants to consent to research, which might somehow entail unacceptable risks and compromise their welfare (Emanuel et al. 2005; Wilkinson et al. 1997). Undue inducement and coercion, associated with vulnerability of research subjects and exposure to risks of harm, therefore constitute perennial concerns for biomedical research, and require elucidation. However, RECs are neither clear nor consistent in their interpretation of undue influence and coercion, and continue, in the absence of clear understanding, to apply what Wilkinson and Moore (1997) term the, "anti-inducement orthodoxy position namely to withhold authorisation" (ibid, p. 375). This chapter aims to examine anew the prohibition against undue influence and coercion in the context of the guidance provided by the law of equity, and to provide practical guidance that can be applied by RECs and those that pursue genetic research in developing countries.

Two case studies

In order to ground the discussion, two examples of apparent undue inducement or coercion are provided, the first case involving an attractive offer combined with extremely high risks for the subject, and the second an imaginary case involving the offer of attractive benefits in order to secure participation of an indigenous community in genetic research.

The first case concerns the well-known Willowbrook hepatitis study, where parents of mentally disabled children, who were the research subjects, provided the informed consent. The parents were encouraged to enlist their children in a study in which they were to be infected with hepatitis (Nelson 1998, pp. 47–66), the inducement being the offer of a guaranteed place for the child in Willowbrook State School, a desirable residential treatment facility for which there was otherwise a long waiting list (Grant and Sugarman 2004, p. 729). An additional inducement to the parents was the promise that the child research subjects would be particularly well treated in the institution (Macklin 1981, p. 3). The parents controversially consented to their children's participation, in exchange for a guarantee of their admission into the Willowbrook State School. As Macklin states, this case was complicated for a number of reasons, including the fact that it involved a highly attractive offer, followed by proxy consent on behalf of vulnerable subjects, who were to be knowingly placed at risk (ibid, p. 3).

The second case is an adapted version of a thought experiment originally suggested by Arnason and van Niekerk (2009). A small tribe in Africa, confined to a few villages but with a strong identity, is found to suffer from an unusually high frequency of a serious disease, which happens to be a common cause of death in developed countries. A group of geneticists expects that a genetic study of the tribe will reveal a single gene, or a fairly small combination of genes, that is responsible for the disease.

Finding the gene, or set of genes, is considered a key to developing a very effective and profitable treatment. Members of the tribe are approached by the geneticists, but they have various concerns about the risks associated with the study, and are not keen to participate. One concern is that they are poor and will never receive access to the resulting treatment. Another is that they might be stigmatised worldwide as the people with this terrible disease. A further concern is that they do not fully understand what unknown risks they might become exposed to via the study. As a tribe they have an unrelated problem, which is that they are being asked to leave their traditional and sacred lands to make way for development initiated by their government and some companies. They need a substantial sum of money to spend on litigation in order to be able to avert the development. The tribe is poor, and needs the money, whilst the geneticists urgently need the bio-samples.

The negotiation or engagement takes place in two phases. Firstly, aware of the tribe's concerns about participating in the study, and also wary of breaching the guidelines, the geneticists make an offer to the participants of a modest but reasonable sum of money to defray participant travel expenses and compensate them for inconvenience, and to the tribal leaders of a modest sum of money to be dedicated to a local health clinic. Whilst some individuals indicate willingness to consent, the tribal leaders refuse to accept the offered terms and effectively prevent the project proceeding. They convey the fact that the risks are unknown, and the rewards offered are simply not sufficiently enticing. After due contemplation by the geneticists, and their commercial partners in the project, of the considerable benefits associated with securing access to the tribe's samples, the geneticists raise their initial offer considerably. They now offer to pay to the tribal leaders, in addition to the normal compensation for expenses and inconvenience, a considerable sum of money, to be used entirely as they wish, and calculated to be an amount sufficient for the tribe to keep the developers away and protect their land.

Even if the tribe's initial concerns about the research risks are real, the improved package of benefits now offered makes it very difficult, and unlikely if not impossible for them to say no under these circumstances. The tribal leaders willingly sign a document signifying their collective permission for the research, informed consent, which sets out all of the relevant details of the research including the terms and conditions relating to the offered payments. Thereafter they encourage the individuals to take part in the study and to give individual consent.

7.2 The Ethical Guidelines

The international biomedical research ethics guidelines have as their explicit purpose the protection of human subjects—individuals and groups—from all forms of exploitation or harm and are an attempt to universalise the moral norms and rules that humans have developed in order to achieve this purpose. Since the Hippocratic Oath, which was written by the Greek scholar Hippocrates in the 5th Century BC who is regarded as the father of Western medicine, physicians and healthcare professionals have sworn to practice medicine honestly and in accordance with a rigorous code of honour (Farnell 2004, p. 244). A succession of modern codes and guidelines has emerged to guide both medicine and research, and which attempt to apply a universal ethical code despite some peoples' concerns about "the undeniable fact of cultural differences" (Tangwa 2007, p. 43).

The international research ethics guidelines are generally regarded as being based upon three core ethical principles first identified and institutionalised in the landmark Belmont Report,[5] namely beneficence, respect for persons, and justice (CIOMS 2002). To these three tenets an important fourth principle has been added, namely that of non-maleficence, which means *do no harm*. Non-maleficence requires a balancing with beneficence in the assessments of risks and benefits (Beauchamp and Childress 2008), and was also said to be a direct response to the Tuskegee and Willowbrook ethical controversies (Hevesi 2010). The guidelines balance a measure of protection of human subjects on the one hand with an emphasis on the autonomy of patients or research participants on the other.

---

[5]The Belmont Report (1979) was adopted by the National Commission for the Protection of Human Subjects in Biomedical and behavioral Research (the National Commission) According to Grant and Sugarman (2004) the report adopted a "principlist" approach in terms of which the three ethical principles of beneficence, respect for persons and justice are *prima facie* binding in consideration of research (p. 274).

The move to autonomy has been suggested to be a departure from the more paternalistic frameworks that had reacted to the early history of coercion of human subjects (Sutrop and Simm 2011, p. 533).

7.2.1 Exploitation in the Guidelines

Ethical concerns about exploitation of both individuals and groups are at the root of the guidelines, which concerns are broadly motivated by the third core ethical principle namely that of justice, and protection of the vulnerable against exploitation. Many terms prevalent in the bioethics discourse, such as consent, voluntariness, risk and inducement have closely mirrored developments in the fields of law concerned with commutative justice or justice in exchange, as discussed in the previous chapter. The securing of informed consent from such vulnerable parties by means of the offering of an excessive, unwarranted, inappropriate or improper reward or other overture (Grant and Sugarman 2004, p. 725), the subject of this chapter, is one of the major bioethical concerns related to the protection of vulnerable participants in biomedical research, particularly in the developing world, as noted earlier.

The ethical principles set out in the Declaration of Helsinki, which revolve to a large degree around autonomy and protection of the human subject, are fleshed out and further amplified in a number of other influential international guidelines. The International Ethical Guidelines for Biomedical Research Involving Human Subjects issued by the Council for International Organisations of Medical Sciences (CIOMS) issued in 1982, and revised first in 1993 and then in 2002, includes amongst its stated priorities informed consent, vulnerability of individuals, groups and communities and populations, and finally equity regarding burdens and benefits (CIOMS 2002, p. 1). The background to the CIOMS introduction describes how the guidelines seek to encourage the avoidance of paternalism by richer countries, and prevention of exploitation of poor countries, whilst guidelines 4–7 prescribe requirements for informed consent. Guideline 5 specifies that the competent individual participant is required, after having considered all the information placed before her to, "arrive at a decision without having been subjected to coercion, undue influence or inducement, or intimidation" (ibid, p. 5).

The common rule of the American Federal Policy for the Protection of Human Subjects provides similar guidance, stating that the prospective subject must be able to consider whether to participate under circumstances that minimize the possibility of coercion or undue influence.[6] Further elucidation of warnings against undue influence is provided by CIOMS in relation to informed consent. Guideline 6, for example, states that sponsors and investigators have a duty to "refrain from unjustified deception, undue influence, or intimidation", the commentary to the guideline correctly reinforcing that "all intimidation invalidates informed consent" (CIOMS, Article 5).

[6]Code of Federal Regulations. Protection of Human Subjects. 45 CFR 46.

These words warn of the influence that a physician, health practitioner or principal investigator in a research project might exert or bring to bear upon a research subject. It should be noted that the forms of prohibited dealings described in the guidelines, namely unfair deception, improper/undue influence, and intimidation are all covered by one of the three legal doctrines against exploitation discussed in the previous chapter. Admitting that the borderline between justifiable persuasion and undue influence is imprecise, and reflecting the debates on the subject in the courts of law, this guideline 6 proceeds to say that the researcher, "should give no unjustifiable assurances about the benefits, risks or inconveniences of the research, for example, or induce a close relative or community leader to influence a prospective subject's decision" (CIOMS 2002, Article 6).

Despite the concerted effort by CIOMS to explain the meaning of undue influence and coercion, it has to be observed that the frequent use of the words *undue, induce and unjustifiable* without any further elucidation or criteria provide little practical guidance to health professionals or laypersons as to the interpretation and precise meanings of these terms. This difficulty resonates with Lord Clyde's conclusion that a precise definition of undue influence is impossible, and that it can be more easily recognised on the facts rather than analysed in the abstract (Royal Bank of Scotland plc v Etridge 2001, p. 1050). Finally, CIOMS guideline 7, which is headed 'inducement to participate', describes how research subjects may be reimbursed for lost earnings and other expenses, as well as being given medical benefits. However, the payments should not be so large or the medical services so extensive as to induce prospective subjects to consent to participate in the research against their better judgement. The commentary to guideline 7 uses phrases such as "against their better judgement" and "rewards that undermine a person's capacity to exercise free choice," in an effort to distinguish and identify the wrongful form of inducement (CIOMS 2002, Article 7).

These explanations reflect the tension between free choice, which is associated with autonomy of persons, and situations where judgement is to some extent and wrongly influenced or clouded, requiring protection. Other guidelines emulate this attempt to describe the prohibition against undue inducement, but are less than helpful, simply repeating words such as influence, improper and inducement. UNESCO's Universal Declaration on Bioethics and Human Rights, for example, simply states "benefits should not constitute improper inducements to participate in research" (UNESCO 2005, p. 15), whilst the Nuffield Council on Bioethics, in its report headed *The Ethics of Research Related to Healthcare in Developing Countries*, describes the nature of undue inducements as those that are able to change a prospective participant's mind with regard to the costs and benefits of the research,

> [...] in a less than benign manner, so that the benefit offered by the inducement *outweighs all risks, however substantial*. This could cause individuals to expose themselves to risks or potential harms that they would *ordinarily view as un*acceptable. And it is in such circumstances that the inducement would be inappropriate. (Nuffield 2002, p. 79) (emphasis added)

Furthermore, the explanation is added that inducements are inappropriate for the reason that they, "jeopardize the voluntariness required for informed consent" (Para 6.2) and might lead a research subject to accept a risk that would "not otherwise be acceptable" (Para 6.29). The guidelines thus resort to circuitous reasoning to describe the situation where the voluntary exercise of choice is trumped by the attractiveness of inducements, which are larger, wrongful or manipulative, using the very words that need elucidation as part of the explanation. Genomic research, whilst fully bound by the ethical principles reflected in the above guidelines, is also guided by additional formulations aimed at dealing with its specific concerns. For example, the HUGO Ethics Committee of the Human Genome Organisation (HUGO) in its seminal statement on benefit sharing stated "that undue inducement through compensation for individual participants, families and populations should be prohibited" (HUGO 2000a, b).

7.2.2 The Central Concerns of the Guidelines

The aforementioned guidelines are the primary guidelines available to RECs entrusted with the task of assessing research projects, and who are required to determine *(inter alia)*, whether or not benefits or inducements offered to research participants are coercive or otherwise undue. The guidance is however, to a large extent circuitous, and is in practical terms according to many authors minimal if not contradictory (Grady 2005; Dickert and Grady 1999). The conclusion of a recent article by Largent et al. captures well the concerns intended to be addressed by this chapter.

> We conclude that human subject's protection professionals hold expansive and inconsistent views about coercion and undue influence that may interfere with the recruitment of research participants and impede valuable research. (Largent et al. 2012, p. 1)

A synbook of the above research guidelines however, enables three interrelated ethical themes to emerge from the pronouncements on undue inducement and coercion. Firstly, it is the vulnerability of certain individuals and groups that demands protection from particular types of exploitative inducements. Secondly, autonomy and voluntariness, essential for informed consent, are jeopardised by undue inducement and coercion; and thirdly, non-maleficence or risk of harm in relation to offered inducements is a central component of this form of potential exploitation, vulnerable people being in danger of consenting to potential harm against their better judgement. The discussion below will examine each of these interrelated ethical themes in turn, bearing in mind the two examples provided above, and will incorporate the guidance on exploitative contracts provided in the previous chapter.

7.3 Vulnerability

Vulnerability demands constant attention in bioethics, being inextricably linked to the idea of exploitation, and recurring throughout any discussion of coercion or undue influence. Vulnerability amongst and within humankind is the subject of a considerable body of research (Forster et al. 2001; Arnason and Schroeder 2013). The word is defined in the International Ethical Guidelines for Biomedical

Research Involving Human Subjects of the Council for International Organisations of Medical Science as follows:

> Vulnerability refers to a substantial incapacity to protect one's own interests owing to such impediments as lack of capability to give informed consent, lack of alternative means of obtaining medical care, or other expensive necessities, or being a junior or subordinate member of a hierarchical group. (CIOMS 2002)

The word spans a vast spectrum of human conditions, ranging from the primary existential fact on the one extreme that as part of humankind all humans share a certain flesh and blood mortal vulnerability, and spreading across multifarious and more obvious criteria and forms of vulnerability—minors, mentally disabled, minorities, the very poor and illiterates to list the most prominent categories—on the other. The commentary to the CIOMS guideline 13 describes vulnerable persons as those who are relatively or absolutely incapable of protecting their own interests, referring more specifically to a lack of power, intelligence, education, resources, strength or other needed attributes (CIOMS 2002).

An analysis of current bioethics guidelines in relation to vulnerability reveals numerous further groups and categories, all of which are deemed deserving of special protection under certain specific circumstances.[7] Article 8 of an earlier version of the Declaration of Helsinki included in a description of vulnerable research populations, those who are economically and medically disadvantaged, who cannot give or refuse consent for themselves, and who may be subject to duress (WMA 2008). This definition cast the conceptual net so widely that much of the developing world could logically have been branded as vulnerable, and was rightly criticised for virtually eliminating vulnerability as a meaningful concept (Forster et al. 2001).

The 2008 version of the Declaration of Helsinki withdrew the previous formulation, choosing to focus on two particular groups, namely "those who cannot give or refuse consent", and "those who may be vulnerable to coercion or undue influence" (WMA 2008). The current version however has simplified the description even further, stating in Paragraph 19, "some groups and individuals are particularly vulnerable, and may have an increased likelihood of being wronged or incurring additional harm. All vulnerable groups should receive specially considered protection" (WMA 2013). The fact that the Declaration has now excluded undue inducement and coercion as two particular forms of vulnerability would appear to represent an advance in thinking, consistent with the views being developed in this chapter. Neither of these formulations however assists with the question as to how to identify such groups, and whether the tendency to exercise poor judgement is in

---

[7]Examples of such further vulnerable groups include racial minorities, the economically disadvantaged, the very sick, the institutionalized, children, prisoners, pregnant women, fetuses, incompetent persons, persons susceptible to coercion and undue influence, junior members of hierarchical groups, subordinate personnel, members of the armed forces, elderly persons, residents of nursing homes, people receiving welfare benefits and other poor people, the unemployed, patients in emergency situations, homeless persons, nomads, refugees or displaced persons, patients with incurable diseases, individuals who are politically powerless, members of communities unfamiliar with modern medical concepts (Hurst 2008, p. 193).

itself a form of vulnerability. If this were the case, all individuals would be potentially vulnerable, at least on occasion, and to that extent in need of protection from their own poor judgement, whilst "impoverished persons might be regarded as automatically vulnerable due to their diminished autonomy" (Macklin 1981, p. 6). The Declaration's new formulation relating to research involving vulnerable groups gives rise to additional concerns.

> Medical research with a vulnerable group is only justified if the research is responsive to the health needs or priorities of this group and the research cannot be carried out in a non-vulnerable group. In addition, this group should stand to benefit from the knowledge, practices or interventions that result from the research. (WMA 2013, Paragraph 20)

Not only does this sentence assume an ability to define such vulnerable groups accurately, but it has been criticised as being potentially unfair as well as patronising in cases where the intended research either had direct health benefits or presented minimal risk for the vulnerable group respectively (Schroeder and Gefenas 2013). The concept of vulnerability is clearly difficult to convey accurately, and will be examined further by first exploring some attempts at definitions, followed by exploring the most extreme examples described as desperate need and low baselines.

7.3.1 Definitions and Categories of Vulnerability

Coleman (2009), in an analysis of vulnerability as a regulatory category in human subject research, has suggested a framework that examines which of the interests of the research participant—individual or group—might be at risk. Firstly, consent-based vulnerability is the person or group's low capacity to protect her interests (ibid, p. 15). Less educated groups, such as the Havasupai tribe, would fall into this category as well as individuals or groups living in dire poverty[8] who are potentially subject to coercion or undue inducement. Secondly, risk-based vulnerability occurs where the person's level of exposure to certain risks is at stake, which would relate to the potential risks associated with the type of research envisaged. Finally, justice-based vulnerability concerns both the distribution of the benefits and burdens of research, which would normally, in cases of genetic research, involve an analysis of whether the research participants were fairly treated, or taken advantage of (ibid, p. 15).

In order to be treated justly, according to Mayer's second class of exploitation, they would need to not only receive fair benefits, but also not be exposed to untoward risks. The justice-based vulnerability of research participants in genetic research, closely associated with the first two forms of vulnerability, well describes the intended emphasis and focus of this book. As stated by Wang (2011), target groups in HPGR are vulnerable in respect of all of these three categories, due to factors such as low education, lack of economic and social resources, cultural sensitivities and an absence of effective legal regulations or guidelines in most developing countries (ibid, p. 117).

[8]What Wertheimer termed a 'low baseline'.

In attempting to arrive at a unifying definition of vulnerability, Schroeder and Gefenas (2009) build on and combine the *external* elements of vulnerability that are contained in most dictionary definitions of the term with the *internal* aspects which comprises the potential failure, otherwise put as incapacity, of certain people or groups to be able to protect their own interests. This distinction is useful in practice, suggesting that policy attempts to address vulnerability, or to reduce the likelihood of harm being suffered, can be approached in two ways, externally by minimising the likely outward causes of harm, and internally by raising the capacity of the vulnerable to protect themselves (ibid, p. 116).

An example of attempts to address these external and internal components of vulnerability is provided by the case of communities that are vulnerable to malaria. External vulnerability is addressed by eliminating the mosquitoes carrying the parasite, and internal vulnerability is addressed by providing the community with knowledge, mosquito nets, repellants and medication. The composite definition of vulnerability proposed by these authors, incorporating both external and internal components, is as follows, "to be vulnerable means to be faced with a significant probability of incurring an identifiable harm, while substantially lacking the ability or means to protect oneself" (ibid, p. 117).

Determining the parameters of who are vulnerable, under what circumstances and in what manner is important for this book. Vulnerability relates directly to the issue of potential exploitation, the identification and prevention of which comprises one of the central tasks of a REC assessing the non-maleficence and beneficence of proposed research. Whilst some research participants might be ostensibly autonomous but suffer from a poor ability to resist a particular form of temptation, others might differ in vulnerability in response to different stimuli. Powerful influences shape peoples' subjective values, choices and decisions, sometimes making them less than fully authentic or independent (Macklin 1981, p. 6). All humans are, to some extent, objectively and subjectively as well as externally and internally vulnerable in differing ways and circumstances. This reality adds to the complexity of the task faced by the biomedical research community, namely to seek to ensure that every research participant or community, even the poorest imaginable, is sufficiently protected in the proposed research, whilst at the same time respecting his, her or their autonomy.

7.3.2 Desperate Need and Low Baselines

Some have suggested *desperate need* as an extreme form of vulnerability that clearly has the potential to reduce voluntariness, and to thereby vitiate consent, to predispose people to take unseemly risks, and to thereby expose themselves to exploitation. It is desperate need that in the words of the ethical guidelines can lead peoples' judgement to be clouded and for them to accede to harmful or risky research "against their better judgement" (Wilkinson and Moore 1997, p. 377). However, as Wilkinson and Moore conclude after examining a number of scenarios, desperate need on the part of a weaker or more vulnerable party on its own, "is not sufficient in itself to undermine consent" (ibid, p. 377). The categories *desperate offeree* and *enormous offer* have been coined in the literature to describe precisely these types of situations, namely where either the vulnerable victim of inducement is in desperate need of money due to poverty, or the amount offered is so huge as to become virtually irresistible (Wilkinson 2005, p. 30). The inference contained in the terminology is that financial incentives above a certain, admittedly difficult to determine level, can make vulnerable or desperate peoples' decisions less autonomous.

Vulnerability of circumstances alone however, does not exclude the right to exercise the most advantageous choice under the circumstances. For example, the choice of a desperately poor person to sell off her property in order to pay for an operation, or to take on dangerous employment in order to reap the rewards. The parents of the mentally disabled children and the tribe in the case studies were all in some degree vulnerable, but their particular states of vulnerability were in none of the cases sufficient cause alone upon which to base a conclusion that autonomy was impaired, or that the consents to participate were impaired and therefore invalid. In addition to their state of vulnerability, what could make such cases abhorrent and deserving of correction, reminiscent of the infamous Tuskegee syphilis trial, is the added factor that the harm was intentionally committed on the unprotected research subjects. This was an exploitative transaction of Mayer's third form, where a transaction took place, which was improper, inauthentic, and simply should not have been allowed.

Following Wilkinson's assertion in regard to cases of desperate need, why indeed should a poor person, parents of a mentally disabled child, or a poor community respectively not be entitled to engage actively and voluntarily in order to improve their dire circumstances? Furthermore, where persons present and hold themselves out as being able to engage without assistance, why should other parties not be entitled to show respect for such persons?

Wertheimer (1996), as an element of enquiry into exchanges, has proposed baselines of individuals, as discussed in the previous chapter. A person's baseline represents the sum total of their circumstances, rights and options prior to the engagement. The holistic enquiry required by equity law would examine first, whether they stood to improve or worsen their baseline, and secondly, in what manner their particular situation was targeted, utilised or manipulated by the other party. In the broad spectrum of human interactions, parties are seldom equal in power and other resources, hence the ethical dilemma as to how exactly to establish when one party to an interaction is so weak as to be categorised as vulnerable in one or other of its multiple potential manifestations.

Godfrey Tangwa, for example, describes research participants in Africa as being often "triple vulnerable" in relation to their research counterparts, in view inter alia, first, of their poverty, second, belonging to medically disadvantaged groups, and in many cases, third, being minors (Tangwa 2009, p. 517). The children in the Willowbrook hepatitis study could similarly have been described as triple vulnerable under those same three categories, and indigenous peoples experience a cultural vulnerability of an altogether different variety. This dilemma of identifying and responding to differing forms of vulnerability is thus likely to remain a perennial concern with regard to the legal rights and obligations that arise from research in developing countries, which places a large responsibility on the REC as the institution charged with approving research.

In conclusion, vulnerability remains a major ethical concern signifying the desire both in bioethics and in law to protect the weak and helpless. Individuals or groups targeted for genetic research can be exposed to one or more of consent-based, risk-based and justice-based vulnerability, and the various groups and types of vulnerability described in the guidelines are not conclusive of the issue. Members of such categories are able under certain circumstances to act with full autonomy, whilst fully autonomous persons can equally on occasion become vulnerable to exploitation. Where individuals belong to a community, broader issues of representation become relevant, and a later discussion will explore particular vulnerabilities of indigenous peoples in relation to genetic research.

On a practical level, the three aspects of vulnerability distinguished in the previous chapter are useful in this analysis, where it plays a different role and is of a different nature in each of the three categories of exploitation. Firstly, people with low socio-economic baselines or in generally vulnerable circumstances are particularly susceptible to manipulation by anybody in a manner regarded as unconscionable dealing.[9] Secondly, a wider group of people in addition to those with low baselines are potentially vulnerable to improper coercion or persuasion by certain persons, who expressly and wrongly apply such pressure on them. This form of dealing is called duress, which is somewhat similar to blackmail, where it is the action of the coercer that contributes to the vulnerability. Thirdly, as a result of a particular relationship of influence or trust, some individuals are particularly vulnerable to the improper or undue influence of such trusted parties.

These three categories of vulnerability relate directly to the legal protections provided by contract law, namely unconscionable dealing, duress, and undue influence respectively. Due to the interrelatedness of baselines, relationships and inducements, whenever issues of vulnerability and exploitation are at stake, evaluations require a comprehensive enquiry into the ability of such participants to act voluntarily in protecting their interests, which leads to the topic of autonomy.

7.4 Autonomy

Autonomy and associated voluntariness, the second of the ethical concerns raised in the guidelines about coercion and undue inducement, are concepts central to the core principle of respect for persons. The principle of autonomy strongly infuses debates not only on vulnerability but also on the moral and legal aspects of informed consent (Rehbock 2011). In this discussion, a general overview on the role of autonomy will be followed by a brief exploration of the limits on voluntariness of consent, and then by an examination of inducements as one of the ways in which informed consent might be influenced and affected. The discussion of inducements will refer to terms used by the guidelines to denote loss of voluntariness, and discuss this under the most

[9]The parents in the Willowbrook hepatitis study, for example.

extreme circumstances provided by the *enormous offer* and the *desperate offeree*. In conclusion, the question will be posed whether and under what circumstances autonomy and thus voluntariness are impaired by exploitative acts or circumstances.

7.4.1 Overview

The notion of individual autonomy as expressed by Isaiah Berlin's statement "I wish my life and decisions to depend on myself, not on external forces of whatever kind" (Berlin 1969, p. 123) is a central principle in bioethics, regarded as an individual's prima facie right to self-determination in regard to her body. This right is closely entwined with the notion of informed consent, the pivotal moral requirement for medical research (Childress and Fletcher 1994, p. 33). Respect for autonomy is generally interpreted as "respect for the autonomous choices of other persons", which has in turn been taken to mean individual liberty and self-determination (Rehbock 2011, p. 524). This understanding of autonomy, namely one that acknowledges the legitimacy of people shaping their lives according to their own aims and values, and which may of course differ between individuals and situations, is thus central to discussions on when such legitimate actions are curbed or tainted by pressures or influences from others.

Another way of depicting autonomy as a principle is succinctly encapsulated by the expression, "persons should be free to choose and act without controlling constraints imposed by others" (Faden and Beauchamp 1986, p. 8). It should be noted that the above individualistic versions of autonomy have been challenged by communitarian philosophers who stress the principle that individuals exist as part of a community (Weijer 1999), as well as by feminist theorists who emphasise the relational nature of all human life (Keller 1997). Another criticism is that this view of autonomy is a particularly Western worldview, and is not shared by other collective or communitarian value systems (Dalton-Brown 2013). These criticisms of the individualistic model have infused debates on consent as well as group exposures to genetic risks, and some authors regard the notion of relational autonomy as being widely accepted with regard in particular to genetic research (Wang 2011, p. 38).

The notion of individual autonomy as expressed by Isaiah Berlin's statement, "I wish my life and decisions to depend on myself, not on external forces of whatever kind" (Berlin 1969, p. 123), is a central principle in bioethics, regarded as an individual's prima facie right to self-determination in regard to her body. This right is closely entwined with the notion of informed consent, the pivotal moral requirement for medical research (Childress and Fletcher 1994, p, 33). Respect for autonomy is generally interpreted as, "respect for the autonomous choices of other persons", which has in turn been taken to mean individual liberty and self-determination (Rehbock 2011 p. 524). This understanding of autonomy, namely one that acknowledges the legitimacy of people shaping their lives according to their own aims and values, and which may of course differ between individuals and situations, is thus central to discussions on when such legitimate actions are curbed or tainted by pressures or influences from others. Another way of depicting autonomy as a principle is succinctly encapsulated by the expression, "persons should be free to choose and act without controlling constraints imposed by others" (Faden and Beauchamp 1986, p. 8).

It should be noted that the above individualistic versions of autonomy have been challenged by communitarian philosophers who stress the principle that individuals exist as part of a community (Weijer 1999, p. 504), as well as by feminist theorists who emphasise the relational nature of all human life (Keller 1997). Another criticism is that this view of autonomy is particularly Western worldview, and is not shared by other collective or communitarian value systems (Dalton-Brown 2013). Beauchamp and Childress (2009) have responded by stating that such relational conceptions of autonomy that derive from social relationships as well as other determinants such as race, class, gender and authority structures are, "illuminating and defensible", as long as they do not obscure the rights of individuals to exercise their own free choice on issues that affect them (ibid, p. 103). These sources of criticism of the individualistic model have infused debates on consent as well as group exposures to genetic risks, and some authors regard the notion of relational autonomy as being widely accepted with regard in particular to genetic research (Wang 2011, p. 38).

As discussed in the previous chapter, the law governing voluntary exchanges—contract law—regards all legal parties as fully autonomous legal persons. This term encompasses all persons, collectives or corporations that are entitled to enter into contracts in their own names, and assumes their personal or collective autonomy. In other words, individuals and collectives are together with corporations deemed to be equally capable of understanding, concluding, and consenting to legally binding agreements. Proper informed consent where the one party is an illiterate individual or a rural community is however, notoriously difficult to achieve and all the more so in exchanges that involve complex issues, for example where the implications and risks associated with genetic research need to be conveyed (Kegley 2004, p. 32).

Informed consent is regarded as the bastion upon which approval for the entire subsequent clinical or research exchange transaction is grounded. A concern exists that in most cases the descriptions of the intended research required for such informed consent are inevitably and in so many cases incomplete (O'Neil 2004, p. 1134). Bioethics and contract law draw strongly upon and influence each other in this regard, as the quality and authenticity of informed consent remains the ethical and legal touchstone of a fair contract between two parties (Berg et al. 2000). Voluntary informed consent, as a central tenet of autonomy, first emerged in the guidelines as a buttress against exploitation during the 1970s, leading to a wealth of ethical analysis as well as a flurry of court cases that fleshed out the moral demands and meaning of the term (Beauchamp 2011, pp. 515–516). The threat of litigation in the form of malpractice suits, in many cases involving analysis of the duty of disclosure of risk that should accompany informed consent, fuelled the refinement both of the legal doctrine of informed consent, as well as of the ethical guidelines (ibid, p. 516). Informed consent has thus evolved as a device to signify autonomous and voluntary acceptance. This begs the question then as to under what circumstances should that voluntariness be limited or curbed?

7.4.2 Limits on Voluntariness of Consent

The question of limits on or the compromising of voluntariness and autonomy is central to all discussions on fairness, equity or exploitation in voluntary exchanges. Voluntariness has been described as a person's right to decide freely, without undue influence, coercion, force or manipulation (Beauchamp and Childress 2009, p. 133). The liberal notion that individuals should be free to enter into all and any transactions coexists with society's acknowledgement that certain limits to such autonomy should be provided. Authors promoting voluntariness in contract, for example, hold that obligations can only be assumed if the terms are meaningfully understood, and the decision is substantially unconstrained (Robertson 2005, p. 180). Systems of law attempt to protect vulnerable groups such as consumers, children, tenants, and research participants, which endeavour is generally regarded as a form of justifiable paternalism (Kimel 2003, p. 118).

The law governing contractual exchanges, as described in the previous chapter, supported both the freedom of as well as the sanctity of contract, in order that commerce might benefit from certainty in exchange transactions. However, in opposition to unbridled freedom of contract, some commentators have asserted that the autonomy of individuals and the liberalistic, individualistic ideals of liberty should be restricted in favour of the public interest, as well as in accordance with the newly compelling ethical frameworks comprised of solidarity and equity (Chadwick et al. 2001).

The assertions of the latter critics are in general that public interest motivations should justify certain compulsory freedom-diminishing actions, such as public vaccinations, the imposition of which might be against the will and autonomy of some. Rehbock (2011), suggests that respect for autonomy should be understood in a broader sense and primarily as respect for the will of the person, bearing in mind that the will is not always the same as the wishes of the person, which may on occasion be immoral, unjust, or even irrational (ibid, p. 526). Her view is that autonomy is to be understood in a Kantian sense more as a moral autonomy rather than as individual liberty. Her explanation being that, where the autonomous will of a person is applied in a manner seen as being immoral or unjust, such a will should indeed be overridden in favour of public interest; and where the application of such autonomous will is irrational, she argues that beneficence should provide the overriding factor (ibid, p. 527). The crucial question raised by the latter viewpoint is who or what institution precisely should make this intrusive judgement to override the expressed wishes of people, and how would appropriate evidence be mustered to support such a radical intervention?

Neumann (2000) captured well the fact that Kant's view of autonomy is radically different from what people might wish for. Kant is not interested in the individual dreams, lives and purposes of human beings, but rather in their functioning as rational, autonomous self-legislators. As he aptly describes the distinction to be made in this approach,

> When I contemplate how to treat you, I'm in no way guided by your natural inclinations, your hopes, desires or dreams. These are put to one side. They are unworthy of you as a person, a *homo noumenon*, and belong to you only as an intrinsically worthless thing, a

> *homo phaenomenon*. Treating your rational nature as an end in itself, I ask whether my actions towards you are consistent with the universal principles of pure practical reason. I ask whether my act could be a universal practice, and willed as such. Once I have done so, I'm through with my moral deliberations: if the act is universalisable, I perform it: otherwise not. No messy consideration of what you want, as a flesh-and-blood human is required, otherwise it is positively excluded. (ibid, p. 286)

According to Neumann then, one does not need to know the unique situations and desires that distinguish persons in order to make universalisable moral decisions. Whilst it is not proposed to explore further the vibrant ethical debate around the place of autonomy and limits on voluntariness (Beauchamp and Childress 2008; O'Neill 2002), moral and ethical conflicts continue to abound in the biomedical context.

Appelbaum et al. (2009), concluded in an empirical study on the voluntariness of consent in research, that voluntariness and its impairment are conceptualised and rooted in the legal doctrine of informed consent (ibid, p. 11), which as the previous chapter described holds that consent is voluntary unless given under some form of improper manipulation, pressure or duress. In an article exploring the limits of voluntariness in contract, Andrew Robertson (2005), states that two different but closely associated claims are dominant in contract law, namely the voluntariness claim and the autonomy claim (Robertson 2005, p. 180), which themes are repeatedly explored in leading judgements, scholarly writings and practitioner treatises.

The former claim embodies the assertion that contractual obligations are voluntary, which can only take place if the terms are meaningfully understood and the decision to enter the contract is substantially unconstrained, whilst the latter claim holds that parties are free to shape their obligations exactly according to their wishes (ibid, p. 180). The distinction between these two is one of emphasis and further dissection is not relevant for this book. Bearing in mind however, the fact that contractual obligations emanate from the will of the parties, it remains the voluntary assumption of obligation that is at the core of all contractual obligations (Craswell 2000, p. 99). Freedom of contract and voluntariness are according to O'Neill synonymous, and in her words, imply no limits as to what inducements could be offered. Classical freedom of contract would permit researchers to offer whatever conditions they like and, "it would be up to subjects whether or not they accepted them" (O'Neill 2002, p. 128).

Limitations on or a reduction of voluntariness are thus synonymous with invalidation of consent. The law of equity brought justice to contract law by intervening wherever the consent of a vulnerable party was somehow improperly obtained. The three forms of exploitative contracts discussed in the previous chapter respond to three distinguishable ways in which the voluntariness of consent is compromised, providing relief in the name of unconscionable dealing, duress or undue influence respectively. In all of these forms of exploitative contracts, in which the baseline of the weaker party would form a factual basis of the enquiry, some form of dealing, including the presentation or offer of inducement would have precipitated the transaction, serving to cloud the mind, coerce or otherwise dominate the better judgement of the vulnerable party. Equity thus provided very useful guidance on how the voluntary consent of a vulnerable person could be impaired. Having discussed autonomy generally, followed by limitations on voluntariness, it is now necessary to examine the inducements that are said on occasion to reduce the voluntariness of their victims.

7.4.3 Inducements

Inducements have long formed an intrinsic part of life, and clearly have an effect on the autonomy or voluntariness of the recipients. In a film titled *Indecent Proposal* featuring Robert Redford as a handsome millionaire, his clearly inductive offer of a million dollars to sleep with a poor and beautiful woman, acted by Demi Moore, and the fraught voluntariness of her subsequent consent become the compelling ethical drama at the heart of the tale. In another commonly referred to example, the consent given by a poor woman for her five children to undergo risky medical experiments in exchange for payment of a large sum of money is examined and generally regarded as morally dubitable (Wilkinson et al. 1997, p. 376). This opens one up to the question as to whether these inducements are ethically wrong, and if so, why?

Inducements and incentives, not necessarily as dramatic as these, are ubiquitous in our lives and are generally employed without too many ethical qualms. China's well-known policy of increased taxation for families with more than one child is an example of a legal disincentive or inducement for citizens to plan for small families, and the punishing by law of speeding car owners provides clear examples of a stern form of official inducement. Inducements can range from benign, to morally dubious, to downright illegal, an extreme example of the latter being pointing a gun at a person's head in order to persuade him to do something. Such inducement for good reason being sanctioned by the criminal law, together with crimes such as bribery, blackmail and extortion.

Negotiation and interaction in the process of forming agreements entails normal and acceptable forms of persuasion but can lead to more obviously exploitative forms of bullying or manipulation that are prohibited both by moral codes as well as by the common law. The negotiation between parties that seek agreement in a transaction is after all, as described by Eisenberg, "a relatively norm-free process centred on the transmutation of underlying bargaining strength into agreement by the exercise of power, horse-trading, threat and bluff" (Eisenberg 1976, p. 638). This less than reassuring quote is reflective of the wry warning provided by Charles Dickens on the dangers and nature of the negotiation process in the 19th Century. "Here's the rule for bargains. 'Do other men, for they would do you.' That's the true business precept" (Dickens 1844, p. 19).

Whilst not all biomedical research is ostensibly commercial in nature, the engagement with potential research participants or communities as part of a genetic research project may nevertheless comprise a form of negotiation, or bargaining with potentially significant consequences. As was discussed in the chapter on law and justice, how to regulate such interactions and to ensure justice in such privately conducted transactions is far from straightforward. Some categories of incentives might raise ethical concerns yet remain difficult to distinguish from legitimate motivation. Such as the practice employed by pharmaceutical companies of rewarding medical doctors in various ways for promoting their drugs, and of acknowledging the achievements of successful researchers with shares, bonuses and other valuable inducements. In a world full of incentives and inducements, the concern of this chapter is that researcher might recruit and incentivise research participants by means that similarly induce them unduly.

The task to determine what makes inducements, regularly used in research, as Grady describes, ethically controversial or contentious in practice (Grady 2005, p. 1681). The carrot and the stick depicted in the well-known metaphor serve as both positive and negative incentives for the donkey's forward motion, and the term *incentive* is used widely and indiscriminatingly as the *carrot* part of the motivation, in some cases as if it were synonymous with the meaning of reward. The word *incentivise* entered the English language meaning to motivate, whereas the word reward, unlike inducement, always implies some form of merit or desert.

For example, a reward offered for information leading to a person's arrest might well motivate a person to come forward, in which case it would serve as an inducement. However, a reward offered to individuals after they have performed feats of bravery might well be fully deserved, but would in this context not have served as a motivator of the particular act, and could therefore not be adjudged as an inducement. The definition of incentives as being, "external prompts to which the individual responds" (Grant and Sugarman 2004, p. 719), thus correctly captures the behaviour-affecting meaning in the word. In attempting to distinguish further what separates incentives from other forms of motivators or rewards offered in negotiations, Grant and Sugarman suggest that the offer must be all of the following:

1. an extrinsic benefit or bonus, (rather than a natural consequence of an act, a reward, or compensation).
2. a discrete prompt expected to elicit a particular response.
3. usually made in the context of an authority relationship, and
4. intentionally designed to alter the status quo by motivating a person to choose differently from how she would choose in its absence (ibid, p. 721).

These four factors could be condensed by simply saying that an incentive is the added element without which the desired action probably would not occur. The guidelines referred to above show that payment of such external prompts, whether as incentives, compensation or honoraria in biomedical research arouse considerable concerns, which led to rules prohibiting coercion and undue influence. Moreover, where research recruitment is carried out by professionals such as doctors, who carry a particular status of respect and trust *vis a vis* potential research participants, controversy surrounds not only potential conflicts of interests, but also the risk of the influencing or coercion of vulnerable participants (Groth 2010).

The above has described incentives generally in research, more importantly we need to address the way in which they are utilised in genomic research. It should be remembered that financial incentives or inducements are distinct from the notion of compensation for time or expenses, often included in research proposals as reimbursement for travel expenses, or even *per diem* payments to compensate for time lost. Compensation and re-imbursement are by definition designed to place the research participant in the position she would have been, and should not contain any of the motivating potential of financial and other incentives.

In practice, incentives and inducements occur at the interface between researchers and communities exchanging information, including the anticipated risks and benefits, related to proposed genomic research. In such an engagement, a community should under normal circumstances always be free to withdraw from the discussions, and research representatives can be expected to employ whatever ethical strategies or powers of persuasion they can muster, including the offering of incentives or related benefits and consequences.

Benefits are not likely to be large in most forms of genomic research, and would generally be of a collective nature such as support for health care or payment into a community trust, but their consideration should not be excluded for the wrong reasons. It is the primary task of institutions and bodies regulating research to abide by the established limits on freedom of contract discussed above, namely to ensure that research approaches are legal, appropriate and do not encroach upon the reaches of ethically permissible behaviour. Relations of power inevitably come into play during such engagements, including a predictable form of asymmetry between researchers and communities in terms of knowledge, resources and wealth, all of which increase the potential for exploitation.

Clearly, tension remains between freedom of contract and autonomy on the one hand, that would allow people to respond freely to inducements, and paternalistic protection of the vulnerable on the other. As will be discussed below, it is the task of the REC to determine the limits of autonomy or voluntariness in respect of any transaction. In order to further explore this particular tension in the context of biomedical research, it will be useful to examine briefly the most extreme challenges possible to the autonomy and informed consent of individuals, namely circumstances where the inducements are extremely large on the one hand, and where the recipient is extremely poor on the other. The underlying question remains whether and when the voluntariness of consent is invalidated.

7.4.4 Extreme Inducements and Needy Offerees

Genetic research is not likely to produce situations where huge benefits are offered to groups of research participants. Too many samples are needed and the benefits available for research communities, if any, are likely to be modest. However, to be completely safe, one can ask the question whether large inducements are wrong. Some authors argue that desperate people with no alternatives, those with a low baseline, or consent-based vulnerability, cannot be other than exploited by extremely attractive incentives (Murphy 1998). Arnason and van Niekerk (2006) have argued persuasively that the size of an inducement itself should not be an ethical problem. Nobody objects, they argue, when a person accepts a huge sum of money to play football for Manchester United, or to perform dangerous stunts for a Hollywood blockbuster.

The real concern is not the size of the offer, but whether the combination of factors, such as vulnerability, relationship, pressure or manipulation of trust, was such as to diminish the autonomy and voluntariness that is required to form a valid and just exchange (ibid, p. 125). According to these authors, a question far more important than the size of the inducement is whether the will of a person has become dominated or otherwise overcome in an exchange. Wilkinson et al. (1997), concur with this conclusion. After examining numerous impecunious individuals' choices to accept potentially harmful but nevertheless financially attractive inducements, they believe that denying any individuals the option of taking inducements, "reduces their freedom, since it removes an option that they clearly prefer to the alternatives" (ibid, p. 377).

Somewhat surprisingly, studies in both developed and developing countries concluded that high inducements do not inevitably lead to impaired voluntariness or poor comprehension of research participants (Pace and Emanuel 2005, p. 11). Other authors have suggested that what is perceived most unfair and objectionable underlying undue inducement concerns is not the nature or level of the incentives offered, as is implied in the *excessive offer* argument, but rather the state of inequality of humankind that leads some desperately poor people to choose differently (Newton 1982, p. 5). In other words, it is regarded as unfair that some people, such as the parents of the mentally disabled children in the first case study above, should find themselves in such pitiful straits, with such low baselines, as to find such dire options attractive.

7.4.5 Is Authentic Consent Possible Under Extreme Conditions?

A question on autonomy of individuals is whether authentic consent is ever possible under the extreme conditions depicted above? There is no doubt that in extreme cases it might be very difficult for a desperate offeree to decline an offer of such attractiveness. However, as concluded by these authors and supported by the previous chapter's discussions on exploitation in contract, neither the size of the incentive nor the poverty of research subjects should automatically vitiate consent. Nor should difficulty of choice be a candidate for reduced autonomy. Some choices require careful weighing of consequences, and individuals have legitimately differing value systems and circumstances. As Macklin (1981), puts it in a response to a critique of her article on undue inducement by Newton (1982), autonomy is difficult to define, is bound to be a matter of degree, and at times the same person exhibits differing degrees of autonomy under different circumstances. In her original article on the subject, Macklin described her thoughts on respect for autonomy as well as acknowledging of the weaknesses of humans and the need for research to protect people against their weaknesses in the following terms,

> [...] respect for persons and their liberty can still recognise the tendency they have to succumb to temptations, to fail to recognise their own best interests on occasion, and to undertake what may be irrational risks... Researchers are not being overly paternalistic in guarding against unduly inducing people to take a risk that only a large sum of money would lure them into incurring. (Macklin 1981, p. 2)

Many authors disagree with the latter part of Macklin's statement. Newton for example, regarded Macklin's concerns for undue inducement as being "insupportable paternalism", whilst stating the real problem to be the fact that the urban poor have so few attractive alternatives for the use of their time (Newton 1982, p. 5).

Others emphasise the duty on society to put basic safeguards in place, prior to protection of autonomy, precisely to assist such vulnerable people (McNeill 1997, p. 390). Under scrutiny therefore, the argument for undue inducement and coercion as cogent reasons not to offer benefits to vulnerable research participants is shown to be difficult to sustain against claims of autonomy, supported by freedom of choice, and objections against paternalism.

In conclusion, this discussion on autonomy has attempted to display its vitality as a primary ethical principle underpinning not only informed consent, but also deeply implicated in assessments of fairness and justice in biomedical research. This book supports the notion expressed by Rehbock (2011) that the autonomy of research participants, no matter how needy, is an important ethical principle, unless it is in some way contrary to the public morals or the public interest (ibid, p. 531). Even in the most unlikely extreme scenario posed for this discussion, involving large payments and/or impoverished research subjects, it may be possible for such persons to cogently defend their right to consent autonomously and to participate voluntarily in the research (Wilkinson 2005, p. 31). Genomic research is not likely to throw up such challenges.

Situations can however, arise where research participants might be manipulated, forced, or otherwise unduly overridden, and the concept of voluntariness remains fundamental to the combating of exploitation within the field of research ethics as within the field of law. The most extreme circumstances postulated above are useful in debating the limits of autonomy of genetic research participants, bearing in mind firstly that inducements in genetic research are likely to be of a more modest nature, and secondly that the interests of genetic research participants are generally situated within a less individual and more relational or collective form.

Respect for persons as a principle and the desire to protect the weak will continue to require a balance in practice, eluding resort to glib formulae. It is suggested that the procedural guidance provided by the law of equity in cases of alleged unfairness in exchange, namely to consider all the relevant procedural factors, most effectively musters the wherewithal for an accurate determination of autonomy and voluntariness. The actual process involved in the transaction: namely the precise manner in which the vulnerable party was approached by the other, bearing in mind the respective baselines, risks, relationships and motives of the parties, enables scrutiny of how the dealing including inducements affected the voluntariness of the other. This provides a tried and tested tool for identifying and distinguishing between different forms of exploitation.

In particular, for voluntariness to be compromised or defective under one of the three forms of exploitative contract, the consent would have to have been obtained either firstly, by means of targeted manipulation focussed on the party's baseline, secondly, by means of some untoward threat or coercive pressure or thirdly, by inducement unduly associated with a relationship of influence or trust. Underlying all such analyses of voluntariness in research is the principle of non-maleficence, and the question to what degree potential participants are to be exposed to risk of harm.

7.5 Risk and Beneficence/Non-maleficence

Genomic research, involving the taking of a DNA sample, does not entail the level of physical risks associated with clinical and other forms of biomedical research. The principle of non-maleficence referred to above and enjoining that no harm should be done to research subjects has become regarded by some as the most important of all the ethical principles (Gillon 1994). This applies equally to genomic research. Risk of harm is the third and final ethical concern extracted from the research ethics guidelines in relation to undue inducement or coercion, the first discussed above being the vulnerability of certain groups and their need for protection, and the second being the autonomy and voluntariness of research participants. Whilst the guidelines accept the need for incentives or inducements in the recruitment process, they warn against their potential to lead research participants to underestimate or to take otherwise unacceptable risks. Discussions around risk are associated with the ethical principles of *beneficence,* connoting acts of kindness, and charity, and is suggestive of altruism, love, humanity, and promoting the good of others (Stanford 2010), as well as the closely associated non-maleficence. The latter, regarded by some authors as the stronger of the two, implies that risks of harm to human subjects as research participants should be minimised, if not avoided. 'First, do no harm' is the translation from the Latin *primum, non nocere* (Beauchamp and Childress 2008).

Application of these central principles determines whether any research project should be ethically permissible, a determination which is prior to and separable from consent (Grant et al. 2004, p. 724). Where risks of harm are not justified by the potential benefits of the research, to research participants, science and the public, research does not pass the tests of non-maleficence and beneficence, and should not proceed. This book will not venture into the rich discussions around these two terms in biomedical ethics (Beauchamp and Childress 2008), but continues directly to the weighing up of risks of harm against benefits, followed by an examination of the particular situation pertaining to genomic research. Thereafter the role and function of RECs in their assessment of research projects is discussed, leading to an examination of some particular and current concerns.

7.5.1 Risks and Harms

Risks are not necessarily bad. Normally functioning autonomous individuals shape their lives to no small degree by the risks that they are prepared to take in the active pursuit of gain. Generally, actions that are associated with known risks are regularly taken in the full knowledge that a lack of success might result, and even on occasion a degree of harm. Humans regularly take risks both in recreation and in the working environment, engaging for example, in risky sports such as skydiving or boxing, and signing up for dangerous careers such as firefighting or the police force fully aware of the attendant hazards. They also legitimately determine and discount what level of risk they are prepared to allow in life.

In fact, the individual regularly and logically trades off the risk of possible future harm in exchange for a more tangible current benefit. Whilst some forms of harm are relatively minor, for example, those that might have embarrassing, annoying, unfortunate or even painful consequences, they might not be deemed sufficiently severe or permanent to undermine what authors have identified as a person's fundamental interests (Macklin 1981; Newton 1982). It is generally agreed however, that the kind of harms that are considered to bear the risk of undue inducement are of a higher magnitude, and should be clearly unreasonable in essence, requiring, "substantial risk of serious physical, psychological, economic, or other harms, which threaten a person's fundamental interests" (Emanuel et al. 2005, p. 337).

The various non-binding ethical guidelines have focussed on ensuring that the informed consent process enables individuals and groups to assess the risks and implications of the proposed research fully, and in addition to dealing with benefit sharing as such, have approved of incentives and provisions for compensating subjects for harms incurred, for time and effort spent, for actual expenses incurred, for inconvenience, as well as for risks taken. Risks of harm have been divided into physical, psychological, social and economic risks (Weijer 2000a, b), and as stated, these aspects of the guidelines have clinical research as a primary focus. So how are these events termed *risks* to be evaluated?

The definition of undue inducement contained in CIOMS guideline 7 warns against the likelihood that an individual's poor judgement might lead to a risk of serious harm (CIOMS 2002, p. 337). As will be explained below, this is not left to the individual to decide, but it is the task of the REC to weigh up whether anticipated risks to participants are of a level to justify the adjective serious. Moreover, whether they are reasonable and acceptable in relation to anticipated benefits, if any, to the participant and to others is also the task of the REC. Determining the seriousness of the risk requires an assessment both of the severity of the possible harm, and of the likelihood of its occurrence.

CIOMS guideline 8 differentiates the standards for benefit-risk assessment for therapeutic and non-therapeutic research, the latter being the subject of this book. Given its focus on therapeutic research, in respect of non-therapeutic research seeking knowledge of potential benefit to humankind, the guidelines state that the risks must be "reasonable in relation to the importance of the knowledge to be gained" (CIOMS 2002, Article 8). A person's fundamental interests, the term used by CIOMS and analogous with the previously described baseline, are not limited to health or economic issues, and can include matters of culture, belief, values and of dignity. According to Grant et al. (2004), where one party deliberately induces religious persons for example, to work on the Sabbath, or to sell their eggs or sperm against their religious scruples, such act would constitute a seductive offer if not a form of bribery resulting in a form of harm (ibid, p. 728).

However, people differ in their convictions no less than in their circumstances. For example, one woman might agree to sell her ova without moral or spiritual compunction, whilst another might equally legitimately refuse due to strongly held cultural or religious convictions. Or a person might volunteer to participate in research for altruistic reasons, despite the fact that this might assail or erode her privacy or dignity within the context of her culture (ibid, p. 730). An example of the latter might be an indigenous person agreeing to be part of a genetic study, despite the fact that there is a risk of possible cultural harm in the form of social stigma or discrimination. Participation in public life, in which commoditisation and economic development inevitably generate a range of different options, therefore and naturally provides continuous challenges to peoples' spiritual and other values (Wilkinson and Moore 1997, p. 122).

7.5.2 Risks in Genomic Research

Genomic research is specifically regulated by a number of international guidelines reflecting the different form of likely risks, including the statements of the HUGO ethics committee referred to above, as well as UNESCO's Universal Declaration on the Human Genome and Human Rights (UNESCO 1997). All of these stress the particular nature of risks relating to identity, privacy and stereotyping that are present. Genome Wide Association (GWA) studies are an example of genetic research, having the potential to extend their impacts far beyond the original research participants, for example, revealing that a stigmatising condition is more likely to occur in one population than another, with obvious social consequences (de Vries et al. 2011, p. 3). As stated by Brody (2002), there is always a danger that communities might inappropriately discount possible long-term harms of such types of genetic research, and that their own judgements can be clouded by short-term benefits (ibid, p. 2856). Genetic research on target groups or communities therefore poses what Tsosie and McGregor (2007), call "distinctive risks and disadvantages" of a collective nature (ibid, p. 352).

Information sourced from individuals provides information relating to the group, which raises issues related to genetic privacy. Such information can lead to potential external harms such as stereotyping or stigmatization, and can also bring about internal harms of a psychological and cultural nature (Tsosie and McGregor 2007). Many of these risks are not only of a less than tangible nature, taking place at some undetermined date in the future, but are in addition often closely associated with group rights linked to the particular culture and ethos of the group (ibid, p. 410). The particular way in which genetic risks are experienced by indigenous and Aboriginal peoples is deserving of special attention, and is discussed in the next chapter.

Whilst it is generally acknowledged that there is less overt risk in contributing towards bio-bank research than in human subject research (Eriksson et al. 2005), various forms of anonymisation are amongst the methods utilised to alleviate the risk of identity-related harm. Nevertheless, the fact that it is difficult to predict or even assess the exact likelihood of such harm, or the attendant risk of harm, is not in any way compensated for by payments made to refund expenses and inconvenience. It raises the question of the fairness of the exchange and whether in Mayer (2007), terms an appropriate and collective requital has been made in relation to the risks undertaken by research participants.

The Declaration of Helsinki provides the fundamental rule that, "every medical research project involving human subjects should be preceded by careful assessment of predictable risks and burdens to the individuals and groups involved in the research in comparison with foreseeable benefits to them and to other individuals or groups" (WMA 2013, p. 17). All the research ethics guidelines similarly require research to be justified on the basis of a favourable risk/benefit assessment (Belmont 1979, Para 46). Whilst protection of the individual and community from untoward risks in research is regarded as paramount, the broader public also has a prior and fundamental interest in the assessment of risks and benefits. As stated in Paragraph 16 of the Declaration of Helsinki, research involving humans should only be conducted if the importance of the objective outweighs the inherent risks and burdens to the research subjects (WMA 2013).

The balancing principle described as the required favourable risk/benefit assessment, first described in the Belmont Report, requires application of what Wertheimer describes as the *reasonable risk criterion* (RRC) (Wertheimer 2011, p. 107). Much academic debate has taken place over the precise weighting of risks and benefits making up this reasonable risk, including risks that are personal or collective, prospective, predictable or potential, present or future, physical or psychological, further analysis of which is not relevant to this study. Only negligible risks to persons are justifiable by the scientific value as well as the social value of research, which according to Emanuel must fully outweigh such risks (Emanuel et al. 2004, p. 934). Risks are thus at the heart of the prior assessment done by a REC, and only those risks deemed necessary to achieve the research objectives are justified.

Emanuel et al. regard two benchmarks as being applicable to developing countries. Firstly the risk-benefit ratio for individuals must be favourable in the context of the affected individuals, and secondly the risk benefit ratio should also be favourable for the broader community (ibid, p. 934). Benefits to be included in the assessment are listed as "information obtained from the study, services provided to participants, or improvements in the health of the community" (ibid, p. 934), and the enquiry remains on the justification or otherwise of the risks to participants. "Only benefits that accrue to participants from the interventions necessary to achieve the research objectives or those deriving from the knowledge to be gained by the research should be used to justify risks to participants" (Freedman et al. 1992, p. 660).

The benefits to society from genomic research are considerable, whilst the benefits to research participants are likely to be small. It is however the attendant risks that perhaps assumes a more central role. A strong espousal of autonomy lies behind the general position, which is that individuals should generally, where the approval of a REC is provided, be allowed to determine the risks that they are prepared to undertake when participating in research. A clear limit on their autonomy or voluntariness is however provided by the interest of the public to ensure that ethically unconscionable risks must not to be incurred, and that vulnerable persons are protected. Hence, the institutional response which ensures that publicly approved institutions (RECs) rather than individuals are required to determine what level of risks are allowed in research.

7.5.3 Assessment by RECs[10]

RECs have been instituted worldwide in order to ensure public accountability by independent review bodies, to safeguard research participants and to provide assurance of non-malificence and beneficence in biomedical research. The Nuffield Council on Bioethics recommended that scientific and ethical review should be undertaken separately, reflecting the twin purposes of the review,

> Research may only be undertaken if the research project has been approved by the competent body after independent examination of its scientific merit, including assessment of the importance of the aim of research, and multidisciplinary review of its ethical acceptability. (Nuffield 2002, Article 7)

The formal and independent evaluation is required to precede the approval of a research project, and only then are individuals invited to participate. This process is designed to provide a sense of assurance to the public wherever such RECs are functioning efficiently and correctly. For example, the infamous Tuskegee and Willowbrook research experiments would have been rejected by properly functioning RECs on the basis of non-maleficence, which decisions would have been quite independent of and prior to any consideration of undue inducement or exploitation. RECs now are bound to carefully assess risks and burdens, and only conduct research "if the importance of the objective outweighs the risks and burdens to the research subjects" (WMA 2013, Paragraph 16).

Primarily scientific experts make the determination as to whether research involving human subjects is scientifically appropriate, and beneficence/non-malificence is assessed by a duly appointed REC. These institutional bodies are required to be not only independent and competent, but also transparent and accountable to the public in order to engender public trust in the ethical conduct of research (Emanuel et al. 2004, p. 934). If the risks of harm for research participants are too high or in other ways unreasonable in relation to the research benefits, it would be deemed unethical to ask anyone to take them, and approval should accordingly be withheld. Bearing in mind the various subjective values, beliefs and preferences of people discussed above, the requirement is that the research should be designed in such a way that the risks are not only clearly discernible, but also both reasonable and necessary to the research. As Coleman puts it, all the risks associated with the proposed research project are required to be worth it from a societal point of view (Coleman 2009 p. 15), as well as being not too risky from an individual point of view. In this regard, it should be noted that societal benefit could never replace a positive risk/benefit ratio for an individual patient/research subject.

What are broadly referred to as the medical and associated risks of participation are not the only subjects of the assessment? Generally, the REC is required to be placed in possession of sufficiently comprehensive information on broader issues

---

[10]As stated earlier, RECs, also called Institutional Review Boards (IRBs) or Ethics Committee (EC) are required to determine the risk-benefit ratio in terms article 17 of the Declaration of Helsinki (2013).

relating to the proposed research, such as the full range of benefits offered to participants, relevant cultural and social issues, the research entity's policy on future utilisation or sharing of findings, and including the possibility of downstream commercial developments or profits (Bauer et al. 2004, p. 114). As has been emphasised throughout this book, justice is a core ethical principle, which applies to the research exchange, the fact that individuals might not receive a fair distribution of the advantages produced by their participation should raise concerns about possible exploitation (Wertheimer 1996, pp. 3–34).

A proper assessment of the entire range of risks and benefits by an REC thus entails analysis not only of the scientific risks or harms directly associated with the research, but also of broader ethical questions, including an evaluation of the entire transaction. At stake ultimately is an overall and holistic assessment of the fairness and equity of the entire research exchange to all parties. However, because the commercialisation of research without compensation is one particular version of potential exploitation and thereby of non-medical risks related to a research project, the presence of this likelihood would thus constitute one of the possible non-medical risks to which research participants are exposed.

Whether financial inducements or benefits to be paid to individuals or a community should be weighed up, as part of the benefits in the risk-benefit assessment is not completely settled. In a paper entitled, *Is payment a benefit?* Wertheimer (2013) disagrees with the standard view, which is that RECs should not regard financial payment as a benefit to participants for the purpose of risk-benefit assessment. He motivates for what he terms an "incorporation view" which holds that RECs should incorporate the value of financial payments as one of the benefits accruing to participants in assessing whether the risks of research are reasonable in relation to the anticipated benefits (ibid, p. 105). He goes on to state that,

> [...] if subjects can reasonably regard the financial benefits of participation as greater than the risks of participation, and if IRBs should demonstrate respect for the interests and judgements of prospective subjects, there is at least prima facie reason for IRBs to incorporate that judgement into their own risk/benefit assessment. (ibid, p. 109)

This book suggests that for genomic research it may be appropriate that the REC should have the flexibility to incorporate and weigh up all forms of benefits for participants, not excluding financial benefits, against the various risks of participation, and in accordance with the non-maleficence and beneficence principle. This process is analogous to the analysis of all relevant facts conducted in the courts of equity described in the previous chapter, where all the circumstances of a case were balanced against principles and rules in a reflexive manner, seeking an equilibrium of equity such as that envisaged by John Rawls. It was clear from the reported cases on contractual exploitation that judges applying the law of equity often differed in such weighing up process, precisely because a single mechanical solution did not exist. It should therefore not be surprising that members of RECs, whilst weighing up risks against the benefits against the principles laid down by the guidelines as well as the moral public policy ethos of the times, might legitimately differ whilst fruitfully debating a fair solution.

In the same article Wertheimer returns to the issue of autonomy of research participants in relation to risk, and concedes the particular difficulty faced by the REC tasked with deciding on whether research with a certain level of risk should proceed. Conceptions of the right thing on the one hand, compete with concerns about "decisional impairments" on the part of the participant on the other (Wertheimer 2011, p. 112). He then endorses a compromise on the issue which he terms a sort of soft paternalism, arguing that the REC should be entitled to override individual decisions based upon what are regarded as cognitive mistakes, whilst rejecting a hard paternalism that would interfere with decisions that the REC thinks are erroneous based on some objective account of a person's interests (ibid, p. 112). Hard paternalism, such as requiring cyclists to wear helmets or motorists to wear seat belts by law, is what some would regard as an extreme form of restriction on freedom of persons. Wertheimer's position is comparable with the view of David Lamb that a principle of *limited paternalism* should be applied in balancing the moral values of health care professionals with the respect for autonomy of patients (Lamb 1995, p. 15).

If, according to this benign paternalism entrusted to the REC a study including the risks of harm and range of benefits is approved, then the general assumption should be warranted that research participants would not be subjected to choices that are harmful, inappropriate, or against their better judgement. In assessing the entire research proposal, the REC is required to ensure not only that the proposed consent documentation adequately reflects all of the facts possibly relevant, but also that prospective participants are to be fully informed of such risk prior to signifying their informed consent. This includes taking into account the broader communal aspects of the affected community, including elders and leaders of extended families which Emanuel describes as the various 'spheres of consent' (Emanuel 2004, p. 934). Engagement with indigenous peoples, a focus of this book, is discussed in the next chapter.[11]

7.5.4 Concerns About Functioning of RECs

Whilst RECs are the institutional pillar upon which ethical principles are assured in research, concerns have been expressed regarding their lack of capacity in developing countries (Kass et al. 2007; Kadam and Karandikar 2012; Ouwe-Missi-Oukem-Boyer et al. 2013). Where such RECs fail, and the concerns of this book focus on indigenous communities targeted for genomic research, the assurance of non-malificence is hollow, and the likelihood of exploitation remains present. Kadam and Karandikar's conclusions relating to RECs in India for example, reflect those of other authors in regard to their functioning and effectiveness in other developing countries, and give cause for concern. As they explain,

> […] many ECs [ethics committees] are oblivious to their roles and responsibilities. It is reported that ECs lack standard operating procedures, do not have a proper composition or adequate representation, thus affecting their functions in regulating clinical research. Moreover, ECs seem to function in isolation, as self-sufficient bodies, having no communication with the regulatory agency or other ECs. (Kadam and Karandikar 2012, p. 50)

[11]Spheres of consent refer to village elders, leaders of extended family, or community. Whilst broader groups provide 'permission', only individuals may validly give consent.

Legislation involving humans in biomedical research in many developing world countries is entirely absent, and RECs in many African states are barely functional (Ouwe-Missi-Oukem-Boyer et al. 2013, p. 10). Even where RECs have been formally constituted, they can nevertheless fall far short of providing the assurance required of them. In addition to the lack of governmental funding and support, other reasons for REC failure include human fallibility resulting in random as well as systemic mistakes, inadequate assessment of cultural and social values, the allowing of excessive risks, and even scientific error (Emanuel et al. 2005, p. 339). The point does not need to be laboured that RECs, particularly in the developing world, do not yet provide the institutional role required of them in research.

Macklin (1981) expressed an additional unease about RECs, pointing out that they are neither qualified nor equipped to conduct intricate investigations into the range of possible subjective value systems, circumstances and vulnerabilities of potential research subjects (ibid, p. 2). This concern is exacerbated in the case of indigenous communities and specific groups targeted for genetic research, who often hold distinct and unique cultural values, and who face a complex suite of potential harms which are not amenable to easy assessment by a group of professionals. In addition, it is surprising in view of the particular difficulties relating to indigenous communities, that few guidelines have been formulated (McMillan and Conlan 2004, p. 206). These particular concerns will be addressed further in the next chapter.

Despite the above concerns, Emanuel joins the authors who believe that the assurance of non-maleficence provided by RECs is institutionally entirely appropriate and fully takes care of the issue of risks and benefits (Emanuel et al. 2005, p. 338). He argues in addition that the excessive concern shown for undue inducements or coercion is generally a mistake, distracting attention from the real problem, which is to improve performance of RECs, "the solution to mistaken assessments of research risks and benefit levels is not to prohibit inducements, but to improve the review system, to focus resources on better training of IRB and ERC members, and better evaluations of research protocols" (ibid, p. 339).

Risk and beneficence/non-maleficence have now been discussed, as the third of the ethical concerns relating to undue inducement and coercion. It was established as a given that some degree of risk is always present in research, and that the prior evaluation of risks juxtaposed with benefits to society as well as to research participants by RECs is the foundation of the non-maleficence and beneficence assessment required prior to the approval of biomedical research. There are many circumstances in which individuals can make themselves vulnerable to unwise choices. However, it is a combination of their baseline circumstances including poverty or need, vulnerability to risk, and the manner in which the inducements are presented, that determines the presence of exploitation and thus wrongfulness in exchange transactions. Whilst individuals as well as communities are generally entitled to exercise their discretion whether or not to accept risks that are associated with genetic research, it is only the non- maleficence assessment of an independent REC that provides assurance both to society as well as to research participants.

If a research study fulfils both the scientific and the ethical requirements, as determined by a properly functioning REC, it can generally be assumed both by the public and by potential research participants that non-maleficence is assured, and that concerns about exploitation of research participants in the form of excessive risks, as well as in the form of undue inducement or coercion, are simply not appropriate (Emanuel 2005, p. 338). Two important concerns have however emerged from this reliance on RECs in relation to risk and exploitation, both of which have relevance for indigenous peoples targeted for genomic research. Firstly, RECs can fail, as noted above. Secondly, genetic research entails collective, rather than individual, interests in relation to an entirely different range of potential external and internal harms, which need to be understood and assessed both in relation to the benefits of the research and to the particular group. These concerns will be addressed in the next chapter. RECs are therefore responsible, bearing in mind the potential for their failure, to carry out prior assessments of both risks and benefits relating to genomic research.

7.6 The Case Studies

Having discussed the three primary ethical concerns relating to coercion and undue inducement, namely vulnerability, autonomy and risk of harm, the two case studies chosen to depict exploitation in biomedical research can now be briefly examined. They were chosen to represent the kind of exploitation feared both in clinical and genomic research respectively. The question addressed in each case was whether, in the light of the ethical concerns discussed above, undue inducement or one of the other two forms of contractual exploitation discussed in the previous chapter were present.

The first example, namely the Willowbrook hepatitis study, throws up the issues of vulnerability and autonomy in stark relief. The mentally disabled child subjects were evidently vulnerable in multiple ways as minors lacking facilities for treatment, and their wellbeing resided entirely in the hands of their parents. They would have fallen clearly under the catch-all phrase used by the Declaration of Helsinki (WMA 2013), namely possibly people who are particularly vulnerable and may have an increased likelihood of being wronged, and Coleman would have referred to them as being both consent-and risk-vulnerable (Coleman 2009). The numerous ethical concerns about this controversial study have been widely debated and are largely beyond the scope of this book.

One can accept with confidence that if a REC had been constituted at the time and had considered this study, it would have been found to be unacceptable on the grounds of non-maleficence, the entire methodology being based upon the intentional commission of harm on the children. No form of benefits, whether scientific benefits of the research, or the social benefits for the children the attractive inducements: offers of residency at Willowbrook State School could have made up for the known risks of intentionally imposed harm associated with the proposed hepatitis infections. The study would thus clearly have been disallowed for on the principle of non-maleficence, namely 'first, do no harm.'

In considering this case, Ruth Macklin concluded that the offer made to the parents was not only an undue inducement, but, based upon her 'moral intuitions,' constituted a coercive offer (Macklin 1981, p. 3). Duress under the law of equity however, which encompasses coercion, requires 'improper pressure or threat of harm' (Bigwood 2003, p. 279). For consent to be coerced, it must be unwilling and to a degree improperly forced. Coercion that involves *legal* pressure however, and as stated earlier, takes advantage of what is termed a pre-existing lack of alternatives in the other party, such as poverty or illness, forms what is termed *lawful duress*, namely a wrong which the courts prefer to distinguish from duress, and to rather define as *unconscionable dealing* (ibid, p. 280).

Macklin's diagnosis of this case as being not only undue inducement, but in fact the more serious abuse of coercion, is therefore not strictly and legally correct. The consent of the parents was not forced, but was eventually willing, following the expressly targeting and manipulation. Again following the guidance provided by the law of equity with regard to the targeted manipulation of a vulnerable party, using a form of coercive pressure which is not illegal to bring about consent, *unconscionable dealing* would be the more accurate legal term for the exploitative transaction. The dealings, namely the offers made to the parents, were thus ostensibly legal, but intentionally exploitative of the parents' particular baseline and form of vulnerability in a manner that the law of equity would punish as being unconscionable; namely dealing that is unfair, inequitable and against public conscience.

In the second example, the tribe described in the thought experiment would have been consent-vulnerable as well as risk-vulnerable, and possibly justice-vulnerable as well as a result of their poverty. Were these vulnerabilities wrongly exploited by the geneticists, and did the second offer constitute exploitation in the form of duress, wrongful coercion or undue inducement?

In considering this case, Ruth Macklin concluded that the offer made to the parents was not only an undue inducement, but based upon her moral intuitions, constituted a coercive offer (Macklin 1981, p. 3). Duress under the law of equity however requires 'improper pressure or threat of harm' (Bigwood 2003, p. 279). For consent to be coerced, it must be unwilling and to a degree improperly forced. Coercion that involves legal pressure however, and as stated earlier, takes advantage of what is termed, "pre-existing lack of alternatives in the other party, such as poverty or illness, forms what is termed lawful duress," namely a wrong which the courts prefer to distinguish from duress, and to rather define as unconscionable dealing (ibid, p. 280). It is therefore felt that Macklin's diagnosis of this case as being not only undue inducement, but in fact the more serious abuse of coercion, is not strictly and legally correct. The consent of the parents was not forced, but was eventually willing, following the expressly targeting and manipulation.

Again following the guidance provided by the law of equity with regard to the targeted manipulation of a vulnerable party, using a form of coercive pressure which is not illegal to bring about consent, unconscionable dealing would be the more accurate legal term for the exploitative transaction. The dealings, namely the offers made to the parents, were thus ostensibly legal, but intentionally exploitative of the parents' particular baseline and form of vulnerability in a manner that the law of equity would punish as being unconscionable; namely dealing that is unfair, inequitable and against public conscience. In the second example, the tribe described in the thought experiment would have been consent-vulnerable as well as risk-vulnerable, and possibly justice-vulnerable as well as a result of their poverty. Did the geneticists wrongly exploit these vulnerabilities, and did the second offer constitute exploitation in the form of duress, wrongful coercion or undue inducement?

The particular form of the tribe's vulnerability first needs to be assessed in relation to all the other factors, which would require answers as to whether the tribe generally was lacking information or alternatives. The questions arise whether the tribe was open to targeted unconscionable dealing by any party, vulnerable to a particular type of coercion or illegal pressure, or vulnerable to inducement by a certain trusted party. All of these factors would have formed part of an examination of the tribe's composite and collective baseline of rights and options, which importantly to note, were pre-existing and not in any way of the researchers' making. Then the effect of the dealings and the relationships would be scrutinised.

The first and second offers of the researchers might well have fallen within the description of what can be termed hard bargaining or even as soft blackmail. However, there is no indication on the facts given that the tribal leaders were unable to understand the risks and benefits, or to engage fully and voluntarily. Whilst the tribe's baseline, its impoverished situation and relative external vulnerability might well have contributed to its lack of alternatives or better options, it chose to exercise the freedom and right to decide for themselves. The tribe's decision to accept the increased offer thus bears all the hallmarks of a voluntary act, rather than one that was manipulated coerced or unduly influenced. The second aspect of the enquiry relating to autonomy, namely whether the consent of the tribe was voluntary, appears similarly to be answerable in the affirmative.

The dealings of the researchers bore no suggestion of exploitative manipulation, nor any manifest or even implied improper threat of harm, the latter, which would have constituted the wrong of coercion. The tribal leaders as well as the geneticists could, at any subsequent juncture, have still chosen to refuse and walk away. Other possible ethical concerns about the transaction, such as whether and to what degree the tribal leaders consulted with the community, and whether their use of the money for the purchase of land rather than for medical purposes was contrary to best practice or guidelines, are issues related to collective representation and decision-making, and not relevant for the issues being addressed in this chapter. Another fact about the dealings was whether a relationship of trust had been created, which predisposed the tribe to accept the offer. This did not appear to be the case.

The final ethical concern to be examined relates to the aspect of risk, which as stated above is a far more complex matter in relation to genetic research. Assuming a REC had approved the study, the distinctive risks and disadvantages described by Tsosie and McGregor (2007), which accompany genetic research and which would potentially affect the entire group rather than only research participants would have been of a complex nature. These would have needed to be carefully assessed in the light of the particular cultural beliefs of the tribe, and failure to do so may have given rise to a collective harm on the entire group.

Proper assessment of such a case study requires assembly of all the above factors in an essentially procedural enquiry similar to that carried out in equity law, to the extent that the information is available. According to Bigwood (2003), a focus on process in the context of contractual justice carefully examines all aspects of the consent, namely whether the parties did in fact authentically agree to the contract (ibid, p. 19). A substantive approach on the other hand would focus purely on the content of the agreement, thus judging the fairness of the outcome against some or other relevant standard (ibid, p. 19). The latter substantive enquiry is reminiscent of Aristotle's justice in exchange, where a bilateral transaction is examined in order to assess whether the outcome is fair and equitable to both, and was applied in the earlier discussions on justice in exchange. On the known facts of this case study, neither of these two approaches yields evidence of wrongdoing or unfairness, as the tribe appears firstly to have validly consented after a robust exchange of positions, and secondly the outcome was not marked by obvious inequity.

The researchers' dealings displayed no targeted manipulation of the tribe, or a coercive offer or implied threat, which would exclude unconscionable dealing and duress (coercion) respectively. Neither did any form of trust or dependency exist between the tribe and the geneticists, which relationship would have been a key requirement for and possible indicator of undue influence. Even if such a relationship of deference or trust had been formed, it would have had to result in the tribe accepting inducements that were in Grant et al. words, "excessive, unwarranted, inappropriate or improper" in order to constitute undue influence (Grant and Sugarman 2004, p. 725). A trusting relationship is in fact to be aspired towards, as will be emphasised in the next chapter on indigenous peoples, but the issue at stake is whether and how such trust may be abused. If for the sake of argument the tribe had in fact formed a trusting relationship with a senior researcher, who had persuaded the leaders that the risks attendant on the genetic research were not to be worried about, their reliance upon such advice rather than upon their own independent knowledge would have made their voluntary but influenced consent to such an agreement a case of undue influence.

As described in the previous chapter, where such a relationship of trust exists, and where undue inducement is alleged, the party benefiting from the transaction would bear the onus or obligation to prove that the vulnerable party acted out of, 'the free exercise of independent will", by showing for example, that such party received independent advice before entering into the transaction (Treitel 1999, p. 380). Undue influence caused by reliance upon an influential and deferential relationship thus remains potentially relevant to all cases where vulnerable peoples or communities engage in relation to genetic research, with insufficient knowledge of the risks, and where their only source of such knowledge is the trusted spokespersons or intermediaries of the research project.

In closing the discussion on this case, the engagement between researchers and tribe bore the hallmarks of two apparently autonomous parties dealing with one another in order to advance their respective interests, and one that would not attract the remedial attention of the law of equity. The eventual agreement carried what appeared to be the tribe's informed consent, free of manipulation, coercion or undue inducement. A nagging concern arises from the tribe's assumption of the future collective risks that are particular to genetic research, which will be addressed specifically in the next chapter. The asymmetrical power and strengths of the two parties as they exchanged offers and incentives and sought to achieve their variously different objectives is not uncommon (Grant 2002, p. 111), but would provide realistic grounds for caution. Nevertheless, it is suggested that a holistic analysis of such exchanges, in which both the process of engagement is examined together with the parties' respective baselines, relationships and dealings, should readily detect the presence of exploitation in one of its forms.

7.7 Summary and Conclusions

This chapter set out to examine undue inducement and coercion as understood today, explaining how they present a serious stumbling block in the way of benefit sharing in genetic research. The origins of the international orthodoxy, warning against payments of benefits to research participants—particularly from the developing world—on the grounds that they might constitute coercion or undue influence, were first sketched, and then amplified by two case studies, which depicted transactions generally deemed to be exploitative for reasons of coercion or undue inducement. Extracts from the research ethics guidelines with jurisdiction over what some scholars have referred to as the legal neighbourhood, that is the stakeholders in genetic research, established the primary ethical concerns. It was also made apparent that the degree of guidance provided to RECs and biomedical researchers on the topics of coercion and undue influence in genomic research is limited. Prevailing ethical concerns as reflected in the guidelines on the topic were found to mirror those issues pivotal in cases of contractual exploitation, namely the protection of vulnerable persons, respect for persons and their autonomy, and risk of harm to participants in relation to inducements. These three inextricably linked considerations were then examined in turn in order to throw light on and understand undue inducement and coercion in genetic research.

Vulnerability as a concept was first explored as an ephemeral element of complex human-ness, which shifts and defies precise definition. Whilst certain groups were variously identified as being vulnerable, the state of being unable to adequately protect one's interests was found to be often circumstantial. The external and internal forms of vulnerability described by Schroeder and Gefenas provided useful guidance for policymakers, whilst Coleman's consent-, risk-, and justice-vulnerability also enabled practical distinctions to be made. The fact that commutative justice is essentially bilateral in nature led to a useful distinction between factual and relational analyses of vulnerability. The former is the primary examination of the person's entire circumstances (baseline) and the latter refers to the person's relationship with the other party. In other words, a person's being potentially vulnerable to a particular approach or dealing by the other person is a central component of the relational analysis that was further developed in the law of equity. Whilst vulnerable people are entitled to and do demand respect for their autonomy, it was noted that fully autonomous individuals are also on occasion susceptible to the making of harmful life choices in response to a particular type of offer, with consequent exploitation.

Autonomy of persons and voluntariness of consent were then explored as the second of the three ethical concerns, calling for appropriate respect for peoples' decisions whilst cognisant of the range of circumstances that can influence their variously activated vulnerabilities. The importance of informed consent as the lynchpin of respect for persons and autonomy was affirmed, with deference to the law of contract, which places voluntary consent as the *sine qua non* of valid legal agreements. Inducements as an intrinsic part of human interactions were then discussed, noting the potential effect of extreme offers on the ability of desperate and vulnerable individuals to exercise valid or authentic voluntary consent.

A number of leading authors firmly espouse the belief that undue inducement and coercion are largely overrated as ethical concerns, their views based upon the assurance provided by an independent REC to approve the scientific merit and non-maleficence of research (Arnason and van Niekerk 2009; Grant and Sugarman 2004). The appropriateness of risks in relation to benefits, they aver in addition, should be the sole domain of the REC, which has the establishment of non-maleficence as a primary purpose. A notable contribution to the analysis of informed consent provided by the law of equity was that the dealing, or the manner in which such consent is obtained from a vulnerable party, differs clearly in each of the forms of exploitative contracts. These three wrongful methods or procedures of inducing consent, namely manipulation, coercion and unduly influencing respectively, all result in a form of consent which is defective, and which is legally addressed by one of the three forms of contractual exploitation.

Risks of harm were finally discussed as the third ethical apprehension relating to undue inducement and coercion in biomedical research. The focus of the ethical principle of non-maleficence was described as being rooted to a large extent in the desire to prevent harm to vulnerable individuals, such as had taken place in infamous scandals such as Tuskegee and Willowbrook. An additional apprehension was however, noted namely that genomic research, which entails minimal physical risks, generates ethical unease of an entirely different nature. Risks relating to dignity, privacy and identity including stigmatization and discrimination are present for all susceptible groups selected for genomic research, including groups defined by race or ethnic identity, geographic situation, disease and other factors. Indigenous communities such as those depicted in the earlier examples of biopiracy, as well as the second case study above present particular challenges in this regard, where lack of trust as well as pervasive differentials of power and of world view constitute more complex collective baselines, and thus present obstacles to genetic research.

The fundamental role and function of the independent RECs required to assess risks and benefits associated with proposed research both from a scientific as well as an ethical point of view was acknowledged as a vital institutional process in research, designed to provide assurance of non-maleficence for research participants and beneficence for society. It was nevertheless emphasised that this assurance falls entirely short whenever RECs are not able to perform this function adequately, a known concern in many areas of the developing world. Where RECs are not able to assess risks and associated cultural concerns of indigenous peoples and other vulnerable groups adequately, the assurance provided by them to avoid exploitation of research subjects cannot be relied upon. However, it would seem more important to focus on putting functioning RECs in place than to place the burden of calculating risk-benefit ratios on individuals without clinical or genetic knowledge.

Finally, the two case studies were discussed in turn. In summary, the Willowbrook hepatitis case was found to be exploitative of the participants primarily due to the fact that the study was in clear breach of the principle of non-maleficence, which would never have been allowed by a properly functioning REC under the research ethics guidelines. Applying the criteria drawn from equity law on the procedural *judicanda*, the form of exploitation was held to be unconscionable dealing, the contractual harm which takes place when a person's particular state of vulnerability—her low baseline—is improperly targeted for the benefit of the stronger party. The second case study involving the tribe selected for genomic research contained all the elements of an asymmetrical exchange, which would normally arouse unease relating to justice and in particular the manner in which the consent of the tribe was obtained. Concerns about whether the risks of harm had been properly assessed, and whether the tribe's fears about the future implications of the genetic research had been duly evaluated by a competent REC, resonated with the general apprehension about genomic research expressed over the past decades by indigenous peoples.

This case highlighted the importance of RECs, juxtaposed with concerns that they are unable to fulfil their essential roles in many developing countries. In order to provide assurance of non-maleficence in genetic research such as the second case study, RECs need to address issues relating to trust as well as to culturally rooted fears of harm, including external harms such as discrimination or stigmatization, and internal harms of a social or psychological nature. The specific concerns of indigenous communities targeted for genetic research are addressed in the final chapter.

In wrapping up the above discussions, it has been shown that benefit sharing should not be prevented or limited by fears of undue influence or coercion in genomic research. In arriving at this conclusion, the following positions have been influential. Firstly, prevention of exploitation of vulnerable research participants, whether by undue influence, coercion, or by the broader form described as unconscionable dealing, remains a legitimate concern in bioethics. However, the identification and scope of these forms of exploitation have been both exaggerated and misunderstood by the research community. Where a REC has approved a research study for non-maleficence and beneficence, benefit sharing should be allowed.

Secondly, the field of genomics was and is not the primary focus of these two forms of protection in research. The type of harms originally intended to be prevented by the prohibition of wrongs such as undue inducement, coercion and unconscionable dealing were primarily physical harms perpetrated on individuals such as the research subjects in the Willowbrook and Tuskegee clinical trials; not the collective, potential and more subtle identity-related harms present in genomic research.

Thirdly, the criteria provided by the law of equity illuminate and guide the accurate identification of these forms of exploitation, by differentiating between the various wrongful forms of dealings or manners in which vulnerable people can be approached and misused in bilateral research encounters. Equity law focussed on the relational aspects of the dealing, identified through the procedural enquiry, in order to categorise the different forms of exploitation.

Fourthly, the RECs primary task is to provide institutional assurance of non-maleficence, which aims to prevent transactions that are inauthentic; the third form of exploitation described by Mayer denoting transactions that are harmful on a vulnerable party and that should not be legal or permissible. A heavy responsibility rests upon RECs as the research community's bastion of assurance against this altogether different form of exploitation. Where RECs fail, for a number of possible reasons, public trust is damaged and genomic research is threatened.

Fifthly and finally, indigenous peoples, representing particular and important forms of communities targeted for genetic research, experience unique forms of risks and challenges that are not currently addressed by adequate guidelines, nor by the presence of functioning RECs, and are addressed in the next chapter.

References

Annas, G., & Grodin, M. A. (1998). Human rights and maternal-fetal HIV transmission prevention trials in Africa. *American Journal of Public Health, 88*, 560–563.

Applebaum, P. S., Lidz, C. W., & Klitzman, R. (2009). Voluntariness of consent to research: A conceptual model. *Hastings Centre Report, 39*(1), 30–39.

Arnason, G., & van Niekerk, A. (2009). Undue fear of inducements in research in developing countries. *Cambridge Quarterly of Healthcare Ethics, 18*(2), 122–129.

Arnason, G., & Schroeder, D. (2013). Exploring central philosophical concepts in benefit sharing: Vulnerability, exploitation and undue inducement. In D. Schroeder & J. Cook (Eds.), *Benefit sharing: From biodiversity to human genetics*. London: Springer.

Arnason, G., & van Niekerk, A. (2006). Undue fear of inducements in research in developing countries. *Cambridge Quarterly of Healthcare Ethics, 18*(2), 122–129.

Belmont Report (1979) *The Belmont Report, Ethical principles and guidelines for the protection of human subjects of research: National commission for the protection of human subjects of biomedical and behavioral research*, Washington, DC, US Government Printing Office: ref 45CFR46. http://ohsr.od.hih.gov/guidelines/belmont.html.

Bauer, K., Taub, S., & Parsi, K. (2004). Ethical issues in tissue banking for research: A brief overview of existing organization policies. *Theoretical Medicine, 25*, 113–142.

Beauchamp, T. L. (2012). Informed consent: Its history, meaning and present challenges. *Cambridge Quarterly of Healthcare Ethics, 20*(4), 515–523.

Beauchamp, T. L., & Childress, J. F. (2008). *Principles of biomedical ethics* (6th ed.). New York: Oxford University Press.

Beauchamp, T. L., & Childress, J. F. (2009). *Principles of biomedical ethics* (6th ed.). New York: Oxford University Press.

Beauchamp, T. L. (2011). Informed consent: Its history, meaning and present challenges. *Cambridge Quarterly of Healthcare Ethics, 20*(4), 515–523.

Berg, J. W., Applebaum, P. S., Lidz, C. W., & Parker, L. S. (eds.). (2000). *Informed consent: Legal theory and clinical practice* (2nd ed.). New York, NY: Oxford University Press.

Berlin, I. (1969). *Two concepts of liberty*. Oxford: Clarendon Press.

Bigwood, R. (2003). *Exploitative contracts*. Oxford: Oxford University Press.

Brody, H. (2002). *The other side of eden: Hunter-gatherers, farmers, and the shaping of the world*. Canada: North Point Press.

Cape Argus. (2011). *Zille's HIV-testing campaign unethical-SAMA*. Staff reporter, (p. 3). Retrieved November 29, 2011, from http://www.iol.co.za/news/south-africa/western-cape/zille-s-hiv-campaign-slammed-by-medical-group-1.1188508.

Craswell, R. (2000). *Against fuller and perdue* (Vol. 67, pp. 99–129). Chicago: University of Chicago Law Review.

CBD. (2010). *Nagoya protocol on access to genetic resources and the fair and equitable sharing of benefits arising from their utilization*. http://www.cbd.int/decision/cop/?id=12267.

Chadwick, R. (2001). Informed consent in medical research. In L. Doyai & J. S. Tobias (Eds.), *British medical journal books* (pp. 203–210). London: British Medical Association.

Chadwick, R., & Berg, K. (2001). New ethical frameworks for genetic databases. *Nature Reviews Genetics, 2*(4), 318–321.

Childress, J. F., & Fletcher, J. C. (1994). Respect for autonomy. *Hastings Centre Report, 24*(3), 33–34.

Chokshi, D. A., & Kwiatkowski, P. (2005). Ethical challenges of genomic epidemiology in developing countries. *Genomics, Society and Policy, 1*(1), 1–15.

CIOMS. (2002). International Ethical Guidelines for biomedical research involving human subjects. *Council for international organisations of medical sciences*. http://www.cioms.ch/publications/layout_guide2002pdf.

Coleman, C. H. (2009). Vulnerability as a regulatory category in human subject research. *Journal of Law, Medicine and Ethics, 37*(1), 12–18.

Dalton-Brown, S. (2013). *Towards a global ethics: the debate on nanotechnology in the European Union and China*. Unpublished Ph.D. book. University of Central Lancashire.

De Vries, J., Bull, S., Doumbo, O., Ibrahim, M., Mercereau-Puijalon, O., Kwiatkowski, D., & Parker, M. (2011). Ethical issues in human genomics research in developing countries. *BMC Medical Ethics, 12*(5), 1–10.

Dickens, C. (1844). *The life and adventures of martin chuzzlewit*. London: Martin and Hall.

Dickert, N., & Grady, C. (1999). What's the price of a research subject? Approaches to payment for research participation. *New England Journal of Medicine, 341*, 198–203.

Eisenberg, M. A. (1976). Private ordering through negotiation: dispute settlement and rulemaking. *Harvard Law Review., 89*, 637–681.

Emanuel, E.J. (Ed.). (2003). *Ethical and regulatory aspects of clinical research: Readings and commentary*. Baltimore, London: The John Hopkins University Press.

Emanuel, E. J., Wendler, D., Killen, J., & Grady, C. (2004). What makes clinical research in developing countries ethical? The benchmarks of ethical research. *Journal of Infectious Diseases, 189*, 930–937.

Emanuel, E. J., Currie, X. E., & Herman, A. (2005). Undue Inducement in clinical research in developing countries: Is it a worry? *The Lancet, 366*, 336–340.

Faden, R., & Beauchamp, T. L. (1986). *A history and theory of informed consent*. Oxford: Oxford University Press.

Farnell, L. R. (2004). *Greek hero cults and ideas of immortality*. Montana: Kessinger Publishing.

Forster, H., Emanuel, E., & Grady, C. (2001). The 2000 revision of the Declaration of Helsinki: A step forward or more confusion? *The Lancet, 358*(9291), 1449–1453.

Freedman, B., Fuks, A., & Weijer, C. (1992). Demarcating research and treatment: A systematic approach for the analysis of the ethics of clinical research. *Clinical Research, 40*, 653–660.

Gillon, R. (1994). Medical ethics: Four principles plus attention to scope. *British Medical Journal, 309*(184).

Grady, C. (2005). Payment of clinical research subjects. *Journal of Clinical Investigation., 115*, 1681–1687.

Grant, R. (2002). The ethics of incentives. Historical origins and contemporary understandings. *Economics and Philosophy, 18*, 111–139.

Grant, R. W., & Sugarman, J. (2004). ethics in human subjects research: Do incentives matter? *Journal of Medicine and Philosophy, 29*(6), 717–738.

Groth, S. (2010). Honorarium or coercion: Use of incentives for participants in clinical research. *Journal of the New York Nurses Association, 41*(1), 11–14.

Hevesi, D. (2010). Robert W McCollum, Dean of Dartmouth Medical School, Dies at 85. The New York Times. Retrieved September 25, 2010, from http://www.nytimes.com/2010/09/26/education/26mccollum.html.

HUGO. (2000a). *HUGO ethics committee. Statement on benefit sharing*. Retrieved April 9, 2000, from http://www.hugointernational.org/Statement_on_Benefit_Sharing.htm.

HUGO. (2000b). *Social, ethical and legal issues committee. Statement on the principled conduct of genetic research*. London: HUGO. http://www.gene.ucl.ac.uk/hugo/conduct.htm.

Hurst, S. A. (2008). Vulnerability in research and health care:describing the elephant in the room? *Bioethics, 22*(4), 191–202.

Kadam, R., & Karandikar, S. (2012). Ethics committees in India: Facing the challenges. *Perspect Clinical Research., 3*(2), 50–56.

Kass, N. E., Hyder, A. A., Ajuwon, A., Appiah-Poku, J. & Barsdorf, N. (2007). The structure and function of research ethics committees in Africa: A case study. *PLoS Medicine, 4*(1) e3. http://www.plosmedicine.org/article/info%3Adoi%2F10.1371%2Fjournal.pmed.0040003.
Kegley, J. A. K. (2004). Challenges to informed consent. *European Molecular Biology Organisation, 5*(9), 832–836.
Kimel, D. (2003). *From promise to contract: Towards a liberal theory of contract*. Oxford: Hart Publishing.
Keller, J. (1997). Autonomy, relationality and feminist ethics. *Hypatia, 12*, 152–164.
Lamb, D. (1995). *Therapy abatement, autonomy and futility: Ethical decisions at the edge of life*. Aldershot, England: Ashgate Publishing.
Largent, E., Grady, C., Miller, F. G., & Wertheimer, A. (2012). Money coercion and undue inducement: Attitudes about payments to research participants. *IRB: Ethics and Human Research, 34*(1), 1–8.
Macklin, R. (1981). Due and undue inducements: On paying money to research subjects. *IRB: A Review of Human Subjects Research, 3*(5), 1–6.
Macklin, R. (2004). *Double standards in medical research in developing countries*. Cambridge: Cambridge University Press.
McMillan, J. R., & Conlon, C. (2004). The Ethics of research related to health care in developing countries. *Journal of Medical Ethics, 30*, 204–206.
McNeill, P. (1997). A response to Wilkinson and Moore: Paying people to participate in research: Why not? *Bioethics, 11*, 390–396.
Murphy, T. E. (1998). Letter. *New England Journal of Medicine, 338*, 838.
Nelson, R. M. (1998). Children as research subjects. In J. P. Kahn, A. M. Mastroianni, & J. Sugarman (Eds.), *Beyond consent: Seeking justice in research* (pp. 47–66). New York: Oxford University Press.
Newton, L. H. (1982). Inducement: Due and otherwise. *IRB Ethics and Human Research, 4*, 4–6.
Nuffield Council on Bioethics. (2002). *The ethics of research related to healthcare in developing countries*. http://www.nuffieldbioethics.org/publications.
Neumann, M. (2000). Did kant respect persons? *Res Publica, 6*, 285–299.
O'Neill, O. (2002). Autonomy and trust in bioethics cambridge, Cambridge University Press, Cambridge.
O'Neill, O. (2004). Informed consent and public health. *Philosophical Transactions of the Royal Society: Biological Sciences., 359*(1447), 1133–1136.
Ouwe-Missi-Oukem-Boyer, O., Munung, N. S., Ntoumi, F., & Tangwa, G. (2013). Capacity building in health research ethics in Central Africa: Key players, current situation and recommendations. *Bioethics Forum, 6*(1), 1–11.
Pace, C., & Emanuel, E. J. (2005). The ethics of research in developing countries: Assessing voluntariness. *The Lancet, 365*, 11–12.
Rehbock, T. (2011). Limits of autonomy in biomedical ethics? conceptual clarifications. *Cambridge Quarterly of Healthcare Ethics, 20*, 524–532.
Robertson, A. (2005). The limits of voluntariness in contract. *Melbourne University Law Review, 29*(1), 179–212.
Royal Bank of Scotland plc v Etridge (no 2). (2001). 3 WLR 1021.
Schroeder, D., & Gefenas, E. (2009). Vulnerability: Too vague and too Broad. *Cambridge Quarterly for Healthcare Ethics, 18*, 113–121.
Schroeder, D., & Gefenas, E. (2013). Realizing benefit sharing: The case for post-study obligations. *Bioethics, 26*(6), 305–314.
Stanford Encyclopaedia of Philosophy. (2010). http://plato.stanford.edu/entries/principle-beneficence.
Sutrop, M., & Simm, K. (2011). Public and private interests in the genomic Era: A pluralist approach in gunning, Holm & Kenway (Eds.) *Ethics law and society surrey*, Aldershot: Ashgate Publishing Limited.
Tangwa, G. B. (2007). How not to compare western scientific medicine with African traditional medicine. *Developing World Bioethics., 7*(1), 41–44.

Tangwa, G. B. (2009). Researh with vulnerable human beings. *Acta Tropica, 9*(112 Supplement), 2–52.

Treitel, G. H. (1999). *Law of contract* (10th ed.). London: Sweet and Maxwell.

Tsosie, R., & McGregor, J. L. (2007). Genome justice: genetics and group rights. *Journal of Law, Medicine and Ethics., 35*(3), 352–363.

UNESCO. (1997). *Universal declaration on the human genome and human rights*. United Nations Educational Scientific and Cultural Organization. http://www.undocuments.net/udhghr.htm.

UNESCO. (2005). *Universal declaration on bioethics and human rights (1997). Human Rights*. United Nations Educational Scientific and Cultural Organization Article 2(f). http://www.unesco.org/new/en/social-and-human-sciences/themes/bioethics/human-genome-and-human-rights/.

Wang, Y. (2011). *Group protection in human population genetic research in developing countries: The People's Republic of China as an example*. Unpublished PhD book. University of Glasgow. http://theses.gla.ac.uk/3005/1/2011WangYuephd.pdf.

Weijer, C. (1999). Protecting communities in research: Philosophical and pragmatic challenges. *Cambridge Quarterly for Healthcare Ethics, 8*, 503–513.

Weijer, C. (2000a). The ethical analysis of risk. *Journal of Law, Medicine and Ethics., 28*, 344–361.

Weijer, C. (2000b). Benefit sharing and other protections for communities in genetic research. *Clinical Genetics* Vol. 58, 5, 367–368.

Wertheimer, A. (1996). *Exploitation*. Princeton University Press: Princeton.

Wertheimer, A. (2011). Is payment a benefit? *Bioethics, 27*(2), 105–116.

Wilkinson, M., & Moore, A. (1997). Inducement in Research. *Bioethics, 11*, 373–389.

Wilkinson, S. (2005). Biomedical research and the exploitation of human tissue. *Genomics, Society and Policy, 1*(1), 27–40.

WMA World Medical Association. (2008). *Declaration of Helsinki: Ethical principles for medical research involving human subjects*. World Medical Association. http://www.wma.net/en/30publications/10policies/b3/.

WMA World Medical Association. (2013). *Declaration of helsinki: Ethical principles for medical research involving human subjects*. http://www.wma.net/en/30publications/10policies/b3/.

Chapter 8
Risks to Indigenous Peoples as Vulnerable Populations

Abstract Indigenous peoples are deserving of special treatment, being not only regularly targeted for genomic research, but also being vulnerable to exploitation in various ways. A description of the peoples and institutions that describe themselves as indigenous or aboriginal is provided, followed by a discussion of the variety of ways in which their world views clash with western notions of an ethical order and of the individual's place in society. The particular forms of vulnerability and risks faced by indigenous populations participating in genomic research are then examined, these being generally associated with identity, and manifesting as discrimination and other collective cultural harms. It is shown that science, claiming to be neutral, is all too often used by dominant cultures in ways harmful to indigenous peoples. Research engagement with indigenous peoples is discussed by first examining various approaches to community consent, as well as a range of specific guidelines that have been formulated over the years to guide western research. Notions of equality, respect, partnership, and community participation in research project are then emphasized, and 'community review' is highlighted as the process recommended to ensure that the voice of the research community remains heard. A Southern African San case is described as an example of ethically problematic genetic research on indigenous peoples, where established research ethics committees failed to ensure adequate protection. Finally, proposals aimed at research ethics committees are provided to improve the current less than satisfactory situation.

Keywords Indigenous peoples · Risks in genetic research · Vulnerable populations · Research ethics · Community review · Community consent · Indigenous community participation in research · Research ethics committee failures · San genetic research

This chapter aims to examine how risk in genetic research is perceived and dealt with by indigenous communities approached to participate in genomic research, with the intention of ensuring the ethical principles *inter alia* of justice and non-maleficence. Justice would be served if, in addition to appropriate benefits being

R.S. Chennells, *Equitable Access to Human Biological Resources in Developing Countries*, DOI 10.1007/978-3-319-19725-8_8

considered in reciprocal exchange for participation, the community were treated as an equal partner in the research, and reasonably assured in respect of potential harms. The discussion on non-maleficence explores the particular type of risk faced by indigenous peoples, and how such risks are best identified and dealt with. The discussion will commence with a brief description of those peoples that identify themselves as indigenous, and then will secondly proceed to answer the question of what risks do they face. Thirdly, the procedural question will be posed, namely how to engage properly with such groups. Finally, the role and capacity of RECs in such research will be examined. The purpose is to add to this book' general conclusions regarding exploitation of communities in the developing world, specific suggestions aimed at ensuring proper inclusion of indigenous populations in genetic research.

8.1 Who Are Indigenous Peoples?

As mentioned in the introductory chapter, certain groups are targeted for genetic research, especially human population genetic research (HPGR) that is also known as population-based genetic research (Wang 2011, p. 10). These groups are normally selected for reasons associated with their particular circumstances, which often include a history of development separate from mainstream society, and includes groups that identify themselves as indigenous, but also using descriptions such as Native Americans, Alaskan Native, Australian Aboriginals, Torres Strait Islanders, and New Zealand Maoris. The National Geographic's Genographic Project, styled as a "landmark quest to decipher our distant past," sought to collect the DNA of "very special people living today" to tell the story of the "human journey" (Wells 2007, p. 45). Spencer Wells, leader of the Project, described these special peoples as follows,

> Ideally, they would be living in the same place as their ancestors did countries ago. They should have been relatively isolated from immigration from surrounding groups who have moved into the region recently. They should also retain some of their ancestor's ways of life, be it language, marriage patterns, or other cultural attributes. In other words, what we want are *indigenous* people. (Italics in original) (ibid, p. 45).

It has been estimated that there are some 5000 distinct indigenous or Aboriginal[1] peoples living in varying degrees of autonomy or assimilation with the larger society (Mander and Tauli-Corpuz 2005, p. 4), and the terms used by authors overlap with other adjectives such as tribal, native, Indian or even local that attempt to describe certain rural communities (Laird 2002, p. 180). The prevailing view is that no formal or universal definition is necessary, but for practical purposes, the

[1]As stated above, the terms indigenous and Aboriginal are synonymous, and are used interchangeably throughout this book.

commonly accepted understanding of the term is that provided in the Jose R. Martinez Cobo's *Study on the Problem of Discrimination against Indigenous Populations*, the working definition of which reads as follows,

> Indigenous communities, peoples and nations are those which, having a historical continuity with pre-invasion and pre-colonial societies that developed on their territories, consider themselves distinct from other sectors of the societies now prevailing on those territories, or parts of them. They form at present non-dominant sectors of society and are determined to preserve, develop and transmit to future generations their ancestral territories, and their ethnic identity, as the basis of their continued existence as peoples, in accordance with their own cultural patterns, social institutions and legal system.[2] (Cobo 2010)

All of the cases of perceived exploitation referred to in the first chapter, including famous cases such as the Hagahai, the Havusapai and the Nuu-Cha-Nulth, consisted of such groups selected because of their particular status and relatively isolated genetic history. McGregor refers to these groups constituted as a result of and for population genomic studies as being socially identifiable groups, despite their use of other defining words such as peoples and communities (McGregor 2007). Group and collective issues are now widely debated in relation to human genetic research, where for example, in genome wide association studies large numbers of participants from selected remote communities are targeted (Greely 2001). Various definitions have been attempted to distinguish groups that choose to be defined as *peoples* (UNESCO 1990), as well as minority groups and ethnic groups (Pogge 1997, p. 193).

HPGR is defined generally in the report of the UNESCO International Bioethics Committee entitled, *Bioethics and Human Population Genetics Research* as, "…a study which aims at understanding the nature and extent of genetic variation among a population or individuals across different groups" (UNESCO 1995) and which research relies upon the large-scale collection of genetic, genealogical and medical data from many individuals (Greely 2000, p. 157). The Human Genome Development Project (HGDP), was one such project, which targeted isolated human populations for the reason that they "contain much more informative genetic records than more recent, urban ones" (Cavalli-Sforza et al. 1991, p. 490).

Such groups can also be referred to as susceptible or vulnerable communities for research purposes, a category which also includes other types of groups such as the Ashkenazi Jews, African Americans, or the very poor (Soskolne 1997). For example, the stigmas faced by the Ashkenazi Jews following genetic research on them has been widely debated (Struewing et al. 1997), raising questions as to who exactly falls into the group, and how are such determinations to be made. Davis (2000) examines groups, communities and contested identities in genetic research, concluding that people can be members of a racial, ethnic or genetically defined group in one of the following ways:

[2][Online] Available at http://www.indigenouspeoples.nl/indigenous-peoples/definition-indigenous.

1. Genetic identity, by carrying genes associated with a group such as Ashkenazi Jews.
2. Ethnic or cultural identity, usually by self-proclaimed affiliation with a community, including issues of language and culture.
3. Religious identity, such as membership in a certain church or synagogue.
4. Stakeholder identity, such as being a parent, spouse or even teacher of a member of a group.
5. Legal identity, such as a recognized membership of a Native American Tribe, and finally.
6. Shared disease experience, (breast cancer survivors) or a shared risk for Huntington's disease (ibid, p. 40).

Whilst this list might not be exhaustive, the point is that genetic research can legitimately separate out a range of such different collective identities or communities, each of which might need to be engaged with in a particular or different manner.

This book examines the problems of indigenous or Aboriginal peoples rather than other types of groups, the former being succinctly described by the following definition used by the indigenous Peoples' Health Research Centre of Canada (IPHRC) report, "indigenous Peoples are the tribal peoples in independent countries whose distinctive identity, values and history distinguishes them from other sections of the national community" (IPHRC 2005, p. 5).

The concept of community and its epistemological underpinnings held generally by indigenous peoples provides new challenges to institutions accustomed to operating within Western notions of an ethical order or the individual's place in society (ibid, p. 6). The IPHRC report expands on this point where it refers to the confrontation of worldviews that exists at the, "meeting point between the dominant Western research paradigm and the emerging Aboriginal epistemological discourse" (ibid, p. 11), which it describes as being most evident when discussing the ethics of matters such as informed consent (ibid, p. 25). In its analysis of the worldview of indigenous peoples, the report states that academics trained in the Western liberal tradition have, "systematically and radically different ideas" from people socialised in Aboriginal societies (ibid, p. 30). Further authors have opposed the domination of the research domain achieved by Western frameworks and worldviews, describing indigenous worldviews and theories as, "challenging the hegemony of Western theoretical production" (Pillai 1999, p. 218).

Indigenous peoples are well on their way to taking ownership over their approach to and participation in genetic research. Organisations such as the Indigenous Peoples Council against Biocolonialism, described above, and the Waitangi Tribunal of New Zealand, a legal commission which published the beliefs and policies guiding bio-prospecting and research on the Maori peoples (Waitangi Tribunal 2011), strongly articulate their particular views and concerns. In Canada, indigenous peoples have lobbied the government for decades, in Australia genetic research with indigenous communities has been the focus of much attention (Kowal 2012), whilst in Africa the interests of indigenous peoples are generally represented by the continent-wide organisation Indigenous Peoples of Africa Coordinating Committee (IPACC). This is a network of 150 indigenous peoples' organisations in 20 African countries. The aims of this organisation are indicative of the aims of other indigenous bodies, and are as follows:

> To promote recognition of and respect for indigenous peoples in Africa
>
> To promote participation of indigenous African peoples in United Nations events and forums
>
> To strengthen leadership and organisational capacity of indigenous civil society in Africa in particular strengthening sub-regional networks of indigenous people.

Each member organisation of IPACC deals with the particular priority issues of identity, rights, recognition and self-determination that affect them in their own country, as are articulated in the UN Declaration on the Rights of Indigenous Peoples (UNDRIP 2007). Amongst the better-known indigenous peoples that belong to IPACC are the Pigmy peoples from Western and Central Africa, the Hadzabe, Masai, Sumburu, Pokot, Barabaig and Okiek from East Africa, the Amazigh, Tuareg and Berbers from West and Northern Africa, and the San and Khoikhoi from Southern Africa.[3] The South African San Council, which represents the San in South Africa has for example, recently responded to deep concerns with regard to the processes and outcomes of genetic research upon them as a community. They requested and held a conference with a group of geneticists, lawyers and ethicists on 9 September 2013, aimed at airing their concerns about a particular genomic research project conducted on Namibian San during 2010, which had caused great unhappiness in the broader San community.

The purpose of the workshop was firstly to address the causes of the mistrust aroused by genomic research where community consultation and consent is perceived as lacking, and secondly, to begin a process of formulating an ethical research protocol to guide all such research in the future. According to the San leaders from three countries[4] that attended the workshop, the San individuals who had provided informed consent for the genomic research were not able to understand and assess the collective risks that the research would pose to the broader community, or the potential damage to be caused by publication of the outcomes. The case study is discussed further below. The apprehensions expressed by the San resonate with Charles Weijer's opinion in this regard, who states that when considering the various philosophical and pragmatic challenges associated with protecting indigenous communities in research, the concerns of such communities are perhaps seen most immediately through the lens of the concept of risk (Weijer 1999, p. 501).

[3]The author is a trustee of IPACC and has acted as the lawyer for the San peoples for the past 15 years. This book does not deal with issues where he would have a conflict of interest as a researcher.

[4]South Africa, Namibia and Botswana.

8.2 What Are the Risks Faced by Indigenous Populations?

What are the specific risks faced by such groups and populations that are targeted for genomic research? How do they perceive and formulate these risks? Unlike clinical research, negligible direct risks exist for individuals who provide DNA samples, whilst the broader community to which one belongs may be put at some present or future risk. A number of legislative and regulatory measures have attempted to address the new risks and challenges arising from genetic research, such as the United States' *Genetic Information Non-discrimination Act of 2008*, which deals inter alia with concerns, that employers or insurers might use genetic information to discriminate against certain groups. Despite such efforts to adjust to the dangers arising from genetic research, it remains a legitimate concern that in much of the world, regulatory standards remain largely focussed on individual rights, and do not yet adequately protect such groups or collectives (Sharp and Foster 2002, p. 145).

Two broad categories of risk exist in genomic research that do not occur in standard medical research, with implications for individuals and groups. Firstly, genetic information discloses information not only about the subject, but also her group, family and community (Annas et al. 1995, p. 360). The publication of a genome sequence taken from a tuft of human hair from an Aboriginal man in the early 1920s resulted in a number of bioethical dilemmas, including questions as to whether and what form of consent should be sought, and if so from which modern group of Aborigines. As stated by Henk Greely, "in a sense, every Aboriginal Australian had something about themselves revealed to the world without their consent" (Callaway 2011, p. 522). Secondly, information about the participant and her relatives or group members has a future component, as a future diary, "with attendant anxiety, discrimination and related concerns" (Annas 1993, p. 2346).

For example, information that a parent has a certain genetic disease, such as Huntington's disease, holds powerful implications for her progeny (Davis 2000, p. 39). Generally, individuals as well as communities approached to participate in genomic research have expressed apprehensions relating to a number of concerns, including privacy (Ginsburg 1999), discrimination and stigmatization (Spaak 2006), commodification and patenting of tissue (Caulfield 2003), amongst other associated harms. These apprehensions are often exacerbated by the different belief systems, and lack of trust of the modern research agenda (Reilly 1998; Kowal 2012; Weijer 1999, 2000b). McGregor has suggested that the risks specific for such socially identifiable populations are identifiable on the one hand as external, which would include discrimination and stigmatization, and then on the other hand as internal which would include dignity, privacy and psychological issues experienced within the group (McGregor 2007, p. 362).

The most commonly perceived external risks are associated with identity and subsequent discrimination or stigmatization based upon the research results (Weijer and Miller 2004, p. 10). Studies involving genetic markers can, for example, reveal that certain participants and their family may be more European than they believed, with social consequences (Sharp and Foster 2002, p. 147). In other cases the identity status of individuals holding political office, or belonging to a certain tribe, might be risked by such genetic testing (ibid, p. 147). Other identity related risks flow from research results that link the group, beyond those that provide samples, with susceptibility to a disease, which might prejudice members in insurance and employment. For example, the risk of society-wide discrimination resulting from published results, the risk of perceptions that a particular indigenous community is, "a drain on the system", and the risks or damage to community cultural identity from results indicating population identity and origin (Weijer and Miller 2004, p. 11). The potential social or psychological impact of the above types of consequences on the affected group, including feelings of shame, humiliation or lack of esteem, is not surprising.

A public example of discrimination resulting from thoughtless publication was provided by the controversy resulting from a geneticist's finding that Maori were twice as likely to carry a gene associated with alcohol and tobacco abuse; the gene was also associated with aggression, and identified as the "warrior gene" (Lee and Chambers 2007). This was widely reported in the media as proving that the Maori were genetically predetermined to be violent, which naturally caused much damage to genetic research in New Zealand (Kowal 2012, p. 19). Maori academics objected strenuously to the racial stereotyping, calling it "unethical and scandalous" (Hook 2009).

McGregor has described these forms of harms as being firstly tangible harms, such as stereotyping or stigmatizing, and secondly dignitary harms, including self-conception, depression, humiliation and the like (McGregor 2007). An example of the latter is Jews perceiving themselves predisposed to cancer, American Indians being prone to alcoholism (ibid, p. 363) or African Americans being associated with sickle-cell disease (Foster et al. 1999, p. 1719). As has been suggested, the overall problem in each case is that members of such groups not even approached to participate or given an opportunity to consent are nevertheless able to be affected or harmed by such research.

It is well known that indigenous communities in many countries have struggled to come to terms with and to prosper in the modern world. Reports of alcoholism, suicide and general dysfunction amongst indigenous peoples abound, leading to a plethora of research and analysis (Royal Commission on Aboriginal Peoples 1996). Some have suggested that unresolved historical grief following trauma and losses across generations have led to the current social problems (Brave Heart and De Bruyn 1998), and other theories postulate historical trauma transmission passed down through memories and story-telling, cultural behaviour, hereditary disposition and psychological processes (Wesley-Esquimaux and Smolewski 2004, p. 76). Whilst some have suggested cultural traits or customs as explanations for the various forms of modern vulnerability, according to Susan Abadian, tribes or peoples subject to social as well as political violence over generations, exacerbated by discriminatory schooling policies, experience a form of trauma which manifests in a state of collective depression (Abadian 1999).

Culture is a broad and multifaceted component of an indigenous or other group identity, which can suffer a form of internal harm following genetic research. Population genetics can for example, have legal implications for a tribe's standing and relationship to land, change perceptions of relationships to other groups, and affect collective notions of ancestry and identity in different ways (Sharp and Foster 2002, p. 147). A community's culture, which includes self-perceptions, spiritual traditions, belief and knowledge systems as well as historical narratives have been shown to be amongst the most sensitive components of genetic research. As described previously, indigenous activist organisations such as the Indigenous Peoples Council against Biocolonialism (IPCB) have stated that the human genetic resources of communities constitute' sovereign cultural property of those communities, which forms part of a legal property regime (Harry and Kanehe 2006).

Other indigenous organisations have made similar ownership claims regarding genetic resources, as did the Mataatua Declaration on cultural and intellectual property rights of indigenous peoples of 1993, which was passed by a plenary of delegates from Ainu (Japan), Australia, Cook islands, Fiji, India, Panama, Peru, Phillippines, Surinam, the United States and Aotearoa (New Zealand) (Mead and Ratuva 2007). These declarations of right have however, not proved effective in securing increased tribal control of samples and data (Reardon and Tallbear 2012, p. 341). Pullman and Nicholas argue strongly against such cultural appropriation by tribes of genetic resources, as well as against placing such cultural property under the dictates and strictures of a property law system. They argue that human DNA need not be treated as exceptional, that no special rules need to be applied, and that research involving DNA should be subject to the common principles of research ethics (Pullman and Nicholas 2011). They warn in addition about the potential of such approaches to conflate the ambiguous and flexible concept of culture with biology (ibid, p. 157).

Cultural harms have been described as the violations of groups' rights to their own culture, (Tsosie 2007, p. 411), which are accepted in International Human Rights Law instruments such as Article 27 of the International Covenant on Civil and Political Rights (UN 1966). For example, migration patterns long held to be sacred parts of indigenous tribes' belief system and culture can be brought into question by genetic research, posing deep problems for the community (Tsosie and McGregor 2007, p. 352). In the Havasupai case, publication of human migration studies resulting from unauthorised research recorded the tribe's origin as having migrated from Asia across the Bering straits, a narrative contradictory to their closely held cultural beliefs on identity and origin (Drabiak-Syed 2010, p. 217).

On being questioned about this research the author stated that it never occurred to her that this theory might be upsetting to the tribe (ibid, p. 219). Markow, the original researcher in this case, defended the publication of the research on human migrations which was objectionable to the tribe, on the grounds that such information was "good science", and of "public benefit" (Hart and Sobraskne 2003, p. 87). A *New York Times* editorial explained that the tribe's deploring of the Bering straits theory was that it ran contrary to their "traditional myths", using terminology that emphasized the journalist's failure to respect or even understand the cultural system of the Havasupai (ibid, p. 219).

Science, in the hands of the dominant culture, claims to be neutral but is used by researchers in a way not always acceptable to indigenous peoples. In the Havasupai case again, the scientists claimed pursuit of knowledge for the public good, but the process proved to be harmful for the tribe. Genetics is increasingly being used as a scientific tool to demonstrate *nativeness,* and the state relies upon genetics in disputes of rights such as land claims in order to determine whether applicants meet a certain "socially constructed image of the Indian" (McCulloch and Wilkins 1995). When the 9000 year old Kennewick Man was discovered in Washington State in 1996, assumptions of his identity as Native American were followed by tribal attempts to claim the remains which were unsuccessful, and the state ordered genetic testing to establish the body's cultural affiliation. When the results were found to be inconclusive, it was ordered to be stored at the Washington University's Burke Museum (Reardon and Tallbear 2012, p. 240). In another case involving ancient ancestral remains, DNA was extracted from a corpse estimated to be have been stored for 10000 years in the ice, which was discovered by the Tlingit tribe of the Prince of Wales Island in Alaska. Two hundred and thirty Native Alaskans voluntarily participated in the study aimed at establishing its identity as well as early population settlement patterns (Kemp 2007).

Agreeing to take part in such studies can, as has been seen many times, lead to unpopular conclusions. The Mashpee Tribe of Massachusetts for example, tried for 32 years to obtain recognition, which was refused after DNA studies revealed a mixed ancestry unsupportive of their identity claim (Pullman and Nicholas 2011). An additional ongoing ethical debate surrounds the question of who has the right to control access to ancient DNA, which may have serious implications on the origins, identity and cultural narrative held by indigenous peoples (Carlyle 2005). The ownership of body parts and DNA stored in museums is generally repatriated to the respective communities, but as stated by Kowal, the British Museum in London generally excludes hair and nails from its repatriation policy, which are a valuable tool for examining human genomes (Callaway 2011, p. 523). The view that genetic knowledge is an objective neutral good that benefits all thus is not equally held or shared by indigenous peoples, for whom the right to control their own culture and identity remains crucial (Reardon and Tallbear 2012, p. 240).

Specific issues related to culture and belief, other than the identity issues discussed above continue to surprise Western researchers. For example, indigenous peoples including the Havasupai often relate to blood in ways that are profoundly spiritual, with reverence, respect and responsibility, and in the same way that they relate to their ancestors (Harry 2009, p. 190). Ahora Mead of the Ngati Awa and Ngati Porou lineage of New Zealand is reported as explaining the sacred nature of blood by saying, "a physical gene is imbued with a life spirit handed down from the ancestors" (ibid, p. 189). Concerning blood and tissue, the National Congress of American Indians stated in a declaration on the issue that, "the taking of blood, hair and tissue samples is an affront to the religious beliefs, cultural values and sensitivities of many indigenous peoples" (ibid, p. 190). Terms such as *Indian blood* and *memory is in the blood* (Olsen 1967) resonate with the burden of an anguished history of colonial oppression, during which the oppressors tried to terminate Indian tribes.

In this quaint poem about a nosebleed entitled, *Cheeky Moon,* author Marie Annharte Baker captures the precarious status of Indian identity and its relationship to blood. It reads,

I'm left to defend,
One lonely drop of blood!
I might terminate,
If I get nosebleed. (Baker 1990, p. 38)

In addition to blood itself, genetic information extracted from the blood too is, from the indigenous perspective, an inalienable part of persons long after it has been removed from the body of an individual (Kowal 2012, p. 9). In the Havasupai case, a scientist later defended his use of DNA samples without consent, saying firstly he had received only transformed cell lines rather than the original tissue, that is descendants of original cells (Hart and Sobraskne 2003, p. 71), and secondly that because the cell lines had no genealogies associated with them, his research was exempt from REC approval (ibid, p. 72). The belief system indicated by this argument gives credence to questions raised by indigenous peoples relating to possible future uses of the samples including their commodification, and to the fair exchange of benefits, if any, resulting there from. In addition, the technical nature of such an argument ignores the real concerns being expressed, comes across as arrogant and dismissive, and is hardly likely to enhance a relationship of trust.

Some authors deny that potential research participants should be warned that one or more findings from the research might challenge their religious beliefs, feeling that this would be too onerous and a burden on science (Reilly 1998, p. 684). Such views support the notion expressed by Markow in the Havasupai case that good science justifies such intrusions. However, after receipt of evidence indicative of the yawning cultural gaps between researcher and community, it is perhaps not surprising that many genetic researchers have reported serious resistance to genetic research involving indigenous and rural communities, one author describing this by simply stating that "indigenous people don't like genetics" (Kowal 2012, p. 19). Kowal also commented that in the lack of clear guidelines as to how to engage with indigenous peoples, declining to participate in research had for many years proved to be appropriate and "the only safe option for Australian Aborigines" (ibid, p. 20).

Richard Grounds describes the eventually negative response of the Yuchi Indian tribe to the approaches of the Human Genome Diversity Project (HGDP), explaining how within most indigenous peoples' epistemologies, the questions of origins, their past and relations to other groups are dealt with in a unique manner

(Grounds 1996). He explains how the caution and mistrust expressed by the tribal elders responded to a history of betrayal and loss perpetrated by the Euro-American society, as follows:

> They are the current expression, the echo, of the long memory of the oppression and genocide perpetrated by Euro-American society upon Native Peoples. The great disparity in levels of trust between Native American and non-Native American societies regarding such a project derive from fundamentally different historical experiences. The issue is not simply the good intentions of the members of the HGDP. (ibid, p. 6)

The gulf between those of European and other descent in countries with a colonial history such as the US, Canada, Australia, New Zealand and parts of Africa has been exacerbated by the advance of science, during which the Western or modern notion of science is often used in a Eurocentric manner to reinforce the belief of what is perceived true and correct. In an article that examines biological anthropology and population genetics, authors Reardon and Tallbear quote a powerful extract from a film, *The Journey of Man,* during which Wells, the population geneticist, tries to level the playing field and engage on an equal plane with an Aboriginal leader whilst providing the following explanation of Western science,

> What I'd like you to think about with the DNA stories we are telling is that they are just that. They are DNA stories. It's our version as Europeans of how the world was populated, and where we all trace back to. That's our song line. We use science to tell us about that because we don't have the sense of direct continuity. Our ancestors didn't pass down the stories. We've lost them, and we have to go out and find them. We use science, which is a European way of looking at the world to do that. You guys don't need that. (Reardon and Tallbear 2012, p. 233)

This passage reflects an earnest attempt by the scientist to engage as an equal, and to acknowledge without judgement the validity of the two differing realities held by his group and the Aboriginals. The authors conclude that most genome scientists and biological anthropologists involved with indigenous DNA miss the deeper histories and meanings of relations between European and indigenous peoples, failing also to understand the deeper and painful connotations of race and culture (ibid, p. 234).

The asymmetry of power between Western and indigenous is omnipresent, and admittedly difficult to deal with in a balanced or satisfactory manner. Rebecca Tsosie examines the social and political nature of research, calling for the development of an "intercultural justice" which would, "restructure the relationships among Native nations and the United States, and alleviate the historical and contemporary grievances and harms that continue to affect native communities" (Tsosie 2007, p. 498). She describes how inherent values and assumptions differ, for example with property, where Euro-American values see all resources as being capable of being owned. Indigenous peoples, on the other hand, understand property to be communal in nature, and recognise that property can have spiritual value (ibid, pp. 397–398).

This radically different worldview, between Western and indigenous Inuit, is described by Hugh Brody in biblical metaphor. He explains that whilst indigenous peoples regard themselves as being of nature, regarding it as perfect in every way, Western society is afflicted by the "curse of Eden" described in the book of Genesis, where God ordered Abraham and his tribes out of paradise, cursed to strive forever to conquer and seek dominion over the birds of the air and the animals of the field (Brody 2001, p. 244). In the light of such potential gulfs of understanding, such vastly disparate worldviews colliding, genetic researchers might well ask how then should such communities be engaged?

8.3 How to Engage with Indigenous Groups?

Engaging properly with such communities for genomic research is no simple matter, the acquiring of proper informed consent being a fundamental requirement. A vast body of research has addressed related questions such as how group or collective informed consent should be secured in such cases (Sharp and Foster 2002). This and the broader debates around genetic exceptionalism, which revolve around the question whether genomic information is so different to other medical information that it requires wholly separate forms of legal protection, will not be engaged with here (Murray 1997; Suter 2001). Whilst clinical research encompasses and envisages various physical harms such as reactions to drugs or procedures, genomic research is normally commenced with a non-invasive sample of DNA from blood, saliva or other tissue, and involves an entirely different and far more subtle form of risk for participants and communities. Some authors have suggested that the extraction of human genomic material has caused the most controversy of all the other issues in indigenous health research (Kowal 2012, p. 19), emphasising the importance of proper engagement aimed at securing authentic informed consent.

8.3.1 Informed Consent

The first of the six charges levelled by the attorneys of the Havasupai tribe was breach of informed consent (Reardon and Tallbear 2012, p. 241). Although this was one of the strongest charges, leading to the eventual settlement, which awarded damages to the tribe, the lessons learned from this case were far broader than informed consent. As can be concluded from the numerous cases referred to earlier including the Havasupai, the Hagahai and the Nuu-Chah-Nulth, where scientists failed to understand the concerns of their research participants, the central problem appears to be that they entirely missed the cultural and worldview context described above. Reardon and Tallbear postulate in this regard that whereas most contemporary genomic researchers reject racism and believe that it has long been abandoned, they nevertheless continue in their methodologies to control native peoples and own their resources (ibid, p. 234).

Pullman and Nicholas (2011), postulate that had either the Havasupai or the Nuu-chah-nulth projects been initiated as an archaeological or anthropological study, the outcomes might have been different (ibid, p. 155). The latter groups and other social scientists, they add, operate within a community-based approach, a research culture that emphasises cultural sensitivity and relationship building. This is in contrast to the impersonal paradigm long established for health research, where research participants are anonymous, and part of a statistically significant sample (ibid, p. 156). Guidelines and proposals for community based research in the fields of science and biodiversity emphasise respect for communities as both users and providers of knowledge, and the shared goals of building healthy communities whilst improving applied scholarship (Bannister 2005; Bannister and Hardison 2006).

The question of how to appropriately obtain informed consent in the community context has spawned a wealth of writings in the field of bioethics, including those that postulate the need for various forms of consultation, permission, and even consent for the group to be required over and above individual informed consent of participants. The issue of different forms of informed consent—broad, narrow, open, general—for population bio-banks and genomic research generally is important for the protection of individuals, but not relevant for the issues of group or collective risks being discussed here (Karlsen et al. 2011; IPHRC 2005). The movement to encompass the collective or communitarian ethos of communities more effectively has nevertheless taken place both in clinical as well as general research, in many cases acknowledging the importance of including community values as part of the design of research (Karlawish 1998).

Regulatory protection was extended from individuals to groups; for example, the National Bioethics Advisory Commission proposed that investigators and RECs should consider how to minimise group harms (NBAC 1999, Article 17). In addition, it proposed that when significant risk to a community can be identified before a study has begun, the researchers should work with community representatives to develop study methods that minimise the potential for harm (ibid, Article 17), and include collective risks as part of the informed consent process (ibid, Article 18). There were those that supported community involvement through dialogue in the development and review of population-specific genetic studies (Greely 1997), whilst others again questioned the viability or practicality of requiring community review (Reilly 1998). A major problem identified with affording such regulatory or veto power to an identified community is the phenomenon described as nesting, where smaller communities exist as part of broader and dispersed or larger communities, and where gate keeping can and does take place (Sharp and Foster 2000, p. 47).

8.3.2 Collective Consent

An article entitled, *Groups as gatekeepers to genomic research: conceptually confusing, morally hazardous, and practically useless,* aptly describes one author's position and concerns about group consent (Juengst 1998). Other authors are concerned particularly about making consultation with communities a requirement as part of regulatory policy, and question whether communities should have the right to actually veto a research proposal (Foster et al. 1999, p. 1720). It should be noted that the CBD and the Nagoya Protocol discussed earlier gave rise to a wealth of commentary on the implementation of prior informed consent as a requirement to ensure fairness for indigenous peoples, much of which gave rise to entirely analogous issues.

Foster et al. agree with those who recommend community dialogue in development and review of research as a policy, for the reason that these authors generally do so on the basis of experiences with populations where culturally specific risks are of collective concern, rather than on the basis of conceptual or academic arguments (ibid, p. 1720). Continuing this trend, Charles Weijer (1999), proposed that a fifth ethical principle, namely respect for communities, should be added to the previously accepted ethical research principles namely autonomy, non-maleficence, beneficence and justice (ibid, p. 505). He suggested that this new principle, effectively summed up by the expression, "be sure and think about the community" is best described as an "emerging duty," namely a *prima facie* duty rather than an absolute duty, which can be set aside if a compelling justification to do so exists (ibid, p. 506). This principle is effectively encapsulated as the placing upon the researcher of an obligation to take the values and choices of the community seriously, and where possible, to protect the community from harm (Weijer and Miller 2004, p. 12).

The need to bridge differing worldviews and to adapt ethical standards formulated within Western paradigms has attracted special attention by governments, such as the Australian government's Commission on Aboriginal Health, as well as the Royal Commission on Aboriginal Peoples that met in 1992 to assist developing an ethical code for research within these communities in Canada (RCAP 1992). Research had acquired a bad name amongst Aboriginal Peoples in Canada, as recounted by Marlene Castanello (2004), elders had complained that they had been badly treated over the years by "misguided and harmful" research, and had according to some been "researched to death" (ibid, p. 98). Reflecting a new mood amongst the assembled Aboriginal leaders that it had become time for them to participate more actively in formulating research ethics codes together with the Western trained researchers, one elder responded, "well, maybe it's time we started researching ourselves back to life" (ibid, p. 98).

Despite the recurrence of "jagged worldviews colliding", the term coined by one leader to describe the encounter between Aboriginal thinking and positivist scientific thought (ibid, p. 103), many of the potentially conflicting issues such as intellectual property rights, informed consent and community control of research agendas were resolved with the formulation of eight principles for an ethics regime for Aboriginal research which are described below. The report drew heavily on research conducted by Charles Weijer, who had examined 16 prior documents on research ethics written by and for Aboriginal Peoples internationally, including Canada, the United States and Australia, which emphasised informed consent from leaders prior to approaching individuals, amongst principles reflecting collaboration with and respect towards the community (Weijer et al. 1999). Generally, consensus was reached on the principle that collective approval if not consent was required for research, which was captured in various documents and guidelines.

8.3.3 Specific Guidelines and Principles for Indigenous Peoples

Despite the striving for fairness in research contained in all the major bioethical guidelines, as well as in the more focussed guidelines aimed at ensuring fair engagement with communities from low and middle income countries promulgated by the Council on Health Research for Development (COHRED 2013), indigenous peoples present particular challenges and complexities which require more explicit approaches. Guidelines specifically for research in indigenous communities have therefore appeared in Australia (NHMRC 1991), Canada (ACUNS 1982; ICC 1992; IPHRC 2005), New Zealand (SPEAR 2008), and the United States (American Indian Law Centre 1994). In addition, a number of institutions have emerged in response to the need to embrace and engage with these communities, such as the Indigenous Peoples' Health Research Centre of Canada (IPHRC 2005), the Indigenous Peoples of Africa Coordinating Committee (IPACC) referred to above, the Indigenous Wellness Research Institute in America (IWRI),[5] and the Australian Institute of Aboriginal and Torres Straits Islander Studies (AIATSIS).[6] The latter body revised its research ethics guidelines in 2010, which focussed primarily on developments in areas such as intellectual property rights and genetic research (AIATSIS 2010; Davis 2000).

The growing international corpus of ethical codes subscribed to by these organisations reflects a number of common features, prominent amongst which is the shared intention to regard and place the relationship between researchers and the community as a form of partnership. For example, full partnership in the research enterprise is now required in all Canadian research ethics guidelines, as originally formulated by the Tri-Council Policy Statement developed by the three major

[5]Indigenous Wellness Research Institute, University of Washington, USA [Online] Available at http://www.iwri.org.

[6]Institute of Aboriginal and Torres Straits Islanders Studies IATSIS. [Online] Available at http://www.iatsis.gov.au.

research granting councils in Canada (TCPS 1998). As a further example of this principle, Australia's National Health and Medical Research Council (NHMRC) establishes equal moral status for both parties, and requires consultation with the community in preparation of the proposal, ensuring sensitivity to the community's culture and politics (Weijer 1999, p. 508).

In addition, these guidelines state that the written informed consent of the community is to be obtained before individuals are approached, and that sensitive issues such as benefits, reimbursement, ownership and disposition of samples and data need to be negotiated during the consultation process. Finally, before publishing of data, the community must be provided with draft copies, and consent must be sought before publication in the media (ibid, p. 508). The most recent version of the NHMRC emphasises the aspiration for ethical relationships, the development of mutual trust and respect, and states unequivocally that, within the research process, "failing to understand differences in values and culture may be a reckless act that jeopardises both the ethics and quality of research" (NHMRC 2003, p. 3).

It would be beyond the remit of this book to describe the numerous guidelines, declarations and documents that have emerged in response to the need to both protect and include indigenous or Aboriginal peoples in research. The following eight abbreviated principles proposed by Marlene Castanello (2004) and based upon a review of current guidelines however provide an accurate reflection of the primary concerns and requirements for such research:

1. Aboriginal peoples have an inherent right to participate as principals or partners in research that generates knowledge affecting their culture, identity and well-being.
2. The Government has a fiduciary obligation to guard against infringement of Aboriginal rights in research activities, which safeguards must be endorsed by Aboriginal representative organisations.
3. Ethical standards of research should strike a balance between restriction and infringement of Aboriginal rights, and respect for the primacy of ethical codes originating in affected communities.
4. Ethical regulation of research affecting Aboriginal peoples should be of broad scope, including protection for all knowledge, languages, territories, material objects, and literary or artistic creations.
5. Aboriginal collective interests in regard to intellectual property should be protected.
6. Aboriginal peoples should develop and implement ethical standards for Aboriginal research, and should serve as majority members on Aboriginal-specific research ethics boards serving local, regional and national communities.
7. The costs of developing and implementing an ethical research regime should be borne by the government.
8. Responsibility for the education of Aboriginal communities should be shared between the communities themselves and the government as well as other organisations. (pp. 111–112)

8.3.4 Guidelines on Consultations and Negotiations

Implementation of institutional arrangements and principles such as those set out above in relation to indigenous communities however, requires a process of engagement, consultation and community review, which in most cases results in some form of negotiations eventually reflected in a contractual arrangement. Negotiations aimed at research contracting are of necessity between research organisations on the one hand and a participant community on the other, arousing concerns about asymmetrical power relations. Such power differentials in bilateral engagements give rise to heightened fears of taking advantage and exploitation of one or other of the forms that have been discussed above, leading to the defences against procedurally unfair taking of advantage developed in the law of equity. The focus on the precise form of the dealing, the improper coercion, or the abuse of trust component of negotiations in equity was an attempt to understand the subtle shifts of power that take place during negotiations.

The Council on Health Research for Development (COHRED) is an example of an international body that has developed guidelines in order to assist low and middle-income country (LMIC) institutions, with the stated purpose of strengthening such LMIC institutions' ability to negotiate fair research contracts with their higher income research partners. The key issues that need to be recorded in order to prevent exploitation and unfairness of the weaker parties include, according to COHRED, intellectual property rights, ownership of data and samples, and related contractual matters. Indigenous peoples as targeted communities within LMICs are similarly at a potential disadvantage *vis a vis* the research institutions, and need assistance over and above the institutional protection provided by society to ensure that guidelines are observed and that their contracts embody the elements of fairness.

Weijer's comparison of sixteen guideline documents written by and for indigenous Peoples, enabled him to extract five essential guiding themes relevant to the process of engagement, aimed at enabling the implementation of Castanello's eight principles described above:

1. Consultation should take place with the community when developing research protocols and throughout research.
2. Informed consent should be obtained from community leaders prior to approaching individuals.
3. Community involvement should take place both in conducting the research and transfer of skills.
4. Access to data and samples, as well as further or additional use of samples should be regulated.
5. Advance drafts of research reports should be distributed to the community to identify community views. (Weijer et al. 1999)

As Weijer and Miller show in a subsequent discussion of protection regimes for various forms of communities in research, community consultation and consent is the highest protection regime, appropriate for highly cohesive communities with legitimate political authorities, which would include indigenous or Aboriginal communities. For other forms of community, including groups such as African American and Ashkenazi Jews, at the very least a process of community consultation is required (Weijer and Miller 2004, p. 13). All of these processes have one common defining feature, namely that members of the study population are respectfully and actively engaged in the process (Sharp and Foster 2000, p. 42).

8.3.5 Group Consent and Community Review

According to Isselmuiden and Faden (1992), a certain consensus has emerged in bioethics in relation to communities involved in research, expressed in the notion that whilst broader groups are able to provide permission, sometimes at the level of consent, only individuals may validly give what is termed informed consent (ibid, p. 832). The group consent contained in all the guidelines on indigenous peoples referred to above reflects the model protocol developed by the North American Regional Committee of the Human Genome Diversity Project (HGDP). The protocol was motivated, in addition to assertions that such groups are generally economically marginal and discriminated against in their own countries, by the fact that, "under such circumstances, it cannot be ethically appropriate to sample some members of a group when the group itself has not agreed to participate" (HGDP 1997, p. 1443).

The outpouring of opposition to the HGDP emphasised that an entirely different approach was required in order to access these valuable sources of genetic information, and that the particular world views of such peoples needed to be engaged with, respected and understood, in order to address their fears and concerns (Dodson and Williamson 1999). The ethics committee of the HGDP proposed that group consent as outlined in their protocol should replace protection of individuals as the dominant paradigm in the management of genetics research with indigenous peoples (Sharp and Foster 2000, p. 41). Group consent raised radical issues, including questions relating to definitions of legitimate groups, and fears of the consequences of such a group veto as mentioned above (Weijer 1999; Davis 2000). Others preferred to suggest the term community permission rather than consent, as a preferable collective process required to accompany individual consent (Diallo et al. 2005, p. 255). Rising above these differences, the term community review emerged and became widely used to describe the careful engagement between researchers and communities in order to address cultural and related concerns.

> Community review thus includes community approval, group consent, communal discourse, and other methods of consulting with communities about the potential implications of genetic research. In its least demanding form, community review could be little more than informal dialogue between researchers and members of the study population. At the other extreme, community review could involve the negotiation of a formal agreement between researchers and the study population. (Sharp and Foster 2000, p. 42).

In some cases the words community approval rather than community consent were intentionally used to describe a less legally rigorous process, such as in the case of a research project involving the Akwesasne Mohawk community in Canada (Sharp and Foster 2002, p. 146). Many subsequent proposals have been made to expand the regulatory system protecting communities from harm, not limited to indigenous communities, and also giving effect to what is now generally regarded as a new ethical principle in genetic research, namely Weijer's "respect for communities" (Weijer 1999, p. 505). Other debates, respond to what one author has termed, "a new era of public interest, controversy and ethical problems" related to "the new human population genetics" (Greely 2001, p. 785). During the course of this debate, the particular need to consult with communities in relation to large-scale genetic databases also received academic attention (Godard et al. 2004). Issues to do with race, ethnicity, disease and other external forms of identifying populations for such research remain relevant and continue to evolve (Davis 2000: Fujimura and Rajagopalan 2011). This book has however elected to focus on the particular types of risks perceived and faced by those described as indigenous communities in genetic research.

8.3.6 The Southern African San Case

A paradigmatic example of reliance upon individual informed consent for genomic research on indigenous peoples, despite the obvious group implications of the research, occurred in a genome-sequencing and extended genetic diversity studies genotyping project published as *Complete Khoisan and Bantu Genomes from Southern Africa* (Shuster et al. 2010). This research project sought to collect and analyse samples from four San, also termed Bushmen in the article, "indigenous hunter gatherers" from Namibia and from one "Bantu" from South Africa (ibid, p. 943).[7] The San were targeted in this research for the reason that they represent, "the oldest known lineage of modern human" (ibid, p. 943), and the stated purpose of the research was to seek a greater understanding of human genetic variation and its effect on human health (ibid, p. 947). The indigenous San of Southern Africa are widely known as descendants of the earliest forms of humankind, and have formed advocacy organisations particularly to protect their interests, and their culture and heritage, since the mid-1990s (Chennells et al. 2009).

The San leaders were unhappy with many aspects of the subsequent publication of the research, which included unreservedly describing the San community as hunter-gatherers. This label however, brought about associated perceptions of stigma and discrimination. Their primary concerns were firstly, the fact that they as the leaders of San had not been consulted, and secondly, the fact that

[7]The Bantu sampled in this project was Archbishop Desmond Tutu from South Africa, a Nobel Peace prize winner and well-known member of the Sotho and Xhosa linguistic groups.

uneducated community members had provided informed consent on this complex project without assistance, each participant being described in the paper as the oldest member of his tribe. A letter addressed to the editor of Nature by San leader Mathambo Ngakaeaja read as follows:

> We were truly shocked when the article was published. None of the official San structures in Namibia had been approached in the customary and expected manner. The Namibian San Council has representatives of all the language groups, and such a project was clearly far too complex to be explained to simple rural San, particularly 'tribal elders' in the words of the article, who were unlikely to have any form of education whatsoever. I can only conclude that no effort was made to contact the community leaders in the haste or alternative secrecy that drove the researchers. (Ngakaeaja 2011)

Numerous further concerns arose from this study. Firstly, ethics approval was provided by three foreign Universities, namely from America, Australia and South Africa, with only a research permit being issued in the host country, Namibia (Schuster et al. 2010, p. 943). Secondly, despite the San being well known and in fact selected for their status as indigenous peoples, none of the established best practices or guidelines referring to research with indigenous communities, such as those described above, were acknowledged or followed. This would have required an entirely different consultative approach to the involvement of the San, giving effect to the ethical principle respect for communities. Thirdly, the individual nature of the informed consent process raised unease not only for the reasons of the complexity of the proposed project and the deliberate selection of illiterate aged participants, but primarily for the fact that the broader San community was never engaged in order to ascertain its inputs or concerns. Fourthly, the project being intended to cover a wide spectrum of issues as is common in genetic diversity studies, the likelihood of research results having the ability to affect, stigmatise or otherwise affect the broader community was predictably high. It is not known whether and in what manner these risks were discussed with the individuals who provided consent.

A document entitled, *Supplementary Information*, published alongside the paper, contained numerous potentially sensitive comments on lifestyle, migration and intermarriage theories relating to the San. These included repeated reference to the low social status of Bushmen, as well as research outcomes relating to genetic markers on personal attributes such as pigmentation, hair colour, lactose persistence, lipid metabolism, bitter taste alleles and hearing ability (Schuster et al. 2010, Supplement pp. 3–7). Finally, the project leaders refused to apologise for their oversight in avoiding community engagement, or to acknowledge any duty to engage with or sign the standard research contract used by the San representative organisation WIMSA.[8]

Despite written requests, the project leaders also failed to provide copies of informed consent films or documents (Begbie-Clench 2010). In response to the

[8]WIMSA (Working Group of Indigenous Minorities in Southern Africa) is a San organisation representing San in Namibia, South Africa and Botswana and based in Windhoek, Namibia. Its website is not currently functional.

request, a member of the research team explained that none of the participants had expressed a wish to be represented by such a San representative organisation, going on to say "as we are dealing with individuals in a personal manner (via their DNA), the individual has the right to participate or not as the information gathered has a direct impact on that person" (Hayes 2010). This view entirely ignored the potential for the collective implications of the study. Whilst the San decided not to take their complaints further, they expressed the fact that the case has left the community deeply mistrustful of genetic research (Ngakaeaja 2011), evoking comparison with Kowals' statement that, "indigenous people don't like genetics" (Kowal 2012, p. 19).

The researchers' attitude in refusing to acknowledge the need to engage with the San leadership on the issue was reminiscent of Markow's defiant defence of her research in the Havusapai case, on the basis that it was good science. The fact that the Havusapai research took place in the 1980s and prior to the evolution of 'respect for communities' as an acknowledged additional ethical principle of research gives pause for thought. This book holds that what some bemoan as the lack of guidelines on how to ensure ethical research with specific indigenous peoples (McMillan and Conlon 2004, p. 206) no longer provides researchers with a legitimate excuse.

In concluding this section on engagement with indigenous peoples as an important category of participants in genomic research, the range of guidelines that have been developed in different countries have a number of principles in common, most important of which is the need to treat such communities with appropriate respect. The principle of respect for communities includes many of the procedural guidelines. For example, the need to engage with them in a participatory process building ethical relationships, and to treat them as equal partners rather than as subservients, must be regarded as a *sine qua non* for modern genomic research. What the San example serves to emphasise however, in addition to a host of ethical failings on the part of the research team, is the fact that RECs from three different developed countries somehow failed to apply essential elements of the above body of guidelines on community engagement. In addition, they failed to give effect to the fifth ethical principle, namely by showing respect for, and by acknowledging the legitimate concerns of the affected San community. RECs were thus shown to be the weak link in the chain of protection, which topic requires further investigation.

8.4 Research Ethics Committees: Roles and Concerns

The San case study shows that RECs, the institutional assurance of non-maleficence, with the responsibility to refer to guidelines that have been developing over the past decades, are by no means reliable forms of protection. Their assessment of non-maleficence is intended to provide the research community with assurance that the research is ethically sound. Issues such as benefit sharing, risk, and the potential for exploitation in the exchange as discussed earlier in this book should be weighed up, and reliably pronounced upon. Genomic research on indigenous communities however presents particular challenges for RECs, as has been demonstrated above, and regulatory guidelines in relation to terms such as undue influence and coercion still reflect the original primary purpose of clinical rather than genomic research. Drawing on the above discussions, the problems with RECs can be regarded as being situated in the realms of attitude, capacity, representivity and process.

8.4.1 Attitude

Attitude can be summed up by the proposed new ethical principle of 'respect for communities' described above, aimed at redressing the individualistic nature of existing moral, policy and legal frameworks (Weijer and Miller 2004, p. 9). RECs should, in the words of Solomon Benatar (2004), ensure that their well-meaning approaches are not paternalistic, meaning that they should be focussed on "helping people from their perspective, and not ours" (ibid, p. 1). The narrow scientific attitude personified by the defiant defence of the benefits of pure medical science given by Markow and others in the Havasupai case, as well as in the failure of Schuster et al. to understand or engage with the grievances expressed by the San leaders, needs to be tackled. The communitarian setting and ethos of genetic research, which often involves vulnerable communities, now requires an attitude that places the status of the participant community as equal to that of the researchers (Weijer 1999). This is well expressed by the call contained in the Australian NHMRC for "ethical relationships" between community and research organisations, including values such as reciprocity, respect, equality, responsibility and integrity (NHMRC 2003). These guidelines and principles are intended to reverse the scientific culture and paradigm reflected in the following explanation by Pullman and Nicholas (2011),

> Because health care workers have been trained in a culture that treats the research subject as a somewhat anonymous participant within a larger research endeavour that purposely set out to separate research from day-to-day clinical practice, they see little need or value in developing relationships with the community, or in reporting results that have no particular relevance to the subject. (ibid, p. 156)

The attitude of REC members can be expected to evolve together with the improvement and refinement of guidelines applicable when indigenous and other vulnerable populations are targeted for research, as well as with training and further development of capacity.

8.4.2 Capacity

It has long been accepted that research ethics in the developing world requires an expansion of the capacity and functioning of RECs in those countries

(Benatar 2000, p. 1137). Despite the consensus on this issue, the pace of improvement appears to be slow. Recent research on research ethics in Central Africa showed that, "legislation governing biomedical research involving humans is absent in almost all the countries in the sub-region," and that many of the countries did not yet have a functioning REC (Ouwe-Missi-Oukem-Boyer et al. 2013, p. 10). Whilst guidelines have been developed in countries where researchers are required to engage with Aboriginal or indigenous peoples, including Australia, Canada and the United States (Weijer 1999, p. 507), the capacity of RECs to engage communities as required in the guidelines is questionable. Calls for increased capacity of developing world RECs that have been made for decades have had limited success, a disturbing fact confirmed by another recent study of genetic research in Africa (De Vries et al. 2012).

This study, after commenting on the lack of capacity of national RECs, confirmed firstly that a clear understanding of the particular research project requires focussed community engagement in securing collaborative agreement of research protocols, and secondly that it is critical to engage with the diverse indigenous, Aboriginal or local communities that are identified for the research (ibid, p. 2). Kowal's description of the general mistrust expressed by indigenous peoples about genetics, de Vries' confirmation of the lack of REC capacity, and the San example of the failure of no less than three RECs to address their most basic cultural interests should conspire to ignite a stern warning against complacency regarding RECs' current capacity to protect.

8.4.3 Representation

As described by Weijer (1999), questions have been asked as to the role of RECs in assessing the representivity of study populations in research protocols. NIH guidelines provide specific mandates in regard representation of minority groups, including whether and to what extent the specific community's interests and views are reflected in the research protocol (ibid, p. 343). RECs should seek the assurance, according to Weijer, that relevant cultural, spiritual and related information is not only made available to them, but also that it is understood prior to their providing institutional approval of the research (ibid, p. 343).

It has been widely acknowledged that outsiders struggle to identify particular culture-based intra-community risks such as those described above, and various proposals have been made to address the issue of representation. These range from suggestions that at least one member of a REC should be a representative of the community, to submissions that indigenous peoples involved in the research should enjoy greater representation, even to form their own RECs. The Indian Health Service (IHS) in America for example, has formed its own REC in order to evaluate research proposals supported by the HIS (Freeman 1998). This specialised REC includes twenty members from American Indian and Alaskan Native communities, including researchers, health professionals and laypersons.

Attempts in Australia to protect the Aboriginal populations have resulted in similar insights in relation to representation. The NHMRC ethical guidelines for example, provide for the creation of specific Aboriginal RECs, and a number of Aboriginal RECs have been established with majority Aboriginal membership. These have the specific brief of reviewing the ethical quality of research proposals involving Aboriginal health (NHMRC 2003, p. 24). The NHMRC guidelines however acknowledge the problem of lack of capacity, namely that not all cases involving Aboriginal and Torres Strait Island people are able to be dealt with by such Aboriginal RECs. They state specifically in response to this reality that non-Aboriginal RECs therefore need to "equip themselves to implement these guidelines" when they encounter such research projects (ibid, p. 24).

This corresponds with Castanello's recommended Principle 6, which states that development and implementation of ethical standards for Aboriginal research should be in the hands of Aboriginal peoples "as majority members on Aboriginal-specific research ethics boards serving local, regional and national communities" (Castanello 2004, p. 111). However, the costs of implementing such an ethical regime have remained an obstacle, as such costs are generally not recognised in the administration budgets of Aboriginal communities and organisations (ibid, p. 111). Given Australia's relative affluence, the problem described above can be expected to be more serious in developing countries.

Representation introduces further challenges. Making sure that ancient cultures are adequately understood is a complex matter, and simply ensuring that one or more indigenous representatives are part of a REC is no guarantee of effective representation of a particular research community's concerns. One of the problems relating to representation has been described as the *nesting* of indigenous communities within larger populations, which share cultures and values to a varying degree (Sharp and Foster 2002). As an example of the nesting problem, individuals who consider themselves Mohawk may do so in a general sense for the reason that they reside in a discrete local community, whilst it is well known that discrete Mohawk communities residing both in the US and Canada have distinct local identities (ibid, p. 47). This means that a Mohawk representative on such a committee would be equipped to provide general guidance regarding Mohawk cultural matters, but would be unable to speak with authority on cultural matters from communities and settlements outside her own personal experience.

The diversity of indigenous or Aboriginal cultures within certain overall shared traits has been repeatedly asserted (Castanello 2004, p. 110). In Southern Africa for example, the broader San community is made up of no less than eight major linguistic groups, in three adjoining countries, who share certain general cultural features and yet differ widely in others (Wynberg and Chennells 2009). Whilst the presence of and consultation with one or more members of a targeted research community is thus regarded as an important component of REC representation, it is rather adherence to a community review process that ensures respect for community to establish trust and satisfy the real concerns of such indigenous peoples.

8.4.4 *Process*

Procedural justice has been described above as being achieved by a legitimate procedure, which ensures that all issues are tabled and all parties are equally heard during the process of a transaction. It was explained that a procedure perceived by all as fair is an absolute requirement for any perception of fairness regarding a distribution of resources (Tyler 1987). Equity law contributed to procedural justice in transactions by ensuring enquiries that focus on all aspects leading up to a transaction, with the spotlight on issues of relationships, motives, power and trust, any of which if misused would lead to identifiable exploitation. Indigenous peoples as well as other groups targeted for HPGR have legitimate concerns both about the burden of risk associated with such research as well as the representation of their concerns and interests on RECs entrusted with their protection. Many of the problems arising from genomic research, epitomised by the legal challenges launched by the Havusapai and Naa-chu-nulth tribes, as well as by the San case and by indigenous peoples' attacks on the HGDP described above, have all arisen from the failure of researchers to engage with the collective concerns of their research participants by means of a fair process. Such failures are seldom intentional, but they have the potential to harm communities as well as to destroy the trust that is needed in order to develop constructive research partnerships.

RECs have been shown to fail to give effect to a respect for communities, to lack capacity in developing world countries, and to experience difficulty in establishing the true nature of cultural and related concerns harboured by indigenous peoples amongst other targeted participant groups. Representation of indigenous peoples on RECs alone has also been shown to provide no guarantee of understanding of specific cultural risks, whilst it nevertheless remains a useful procedural component when combined with the application of sound policies relating to community participation. Amongst the ethical guidelines developed to protect indigenous communities in genomic research which should be observed, the community review process described by Sharp and Foster above has the potential to fulfil this need. If followed with sensitivity to the diverse forms of research, risks and communities, community review is a process that provides considerable assurance both to the REC as well as to the community of procedural justice and ethical engagement.

8.5 Conclusion and Proposals

The purpose of this chapter was to address the third research question of the book, namely whether the particular concerns of indigenous peoples are adequately met by the current research practices and guidelines.

The purpose of this chapter was to address the third research question of the thesis, namely whether the particular concerns of indigenous peoples are adequately met by the current research practices and guidelines. The previous chapter, whilst examining two particular forms of procedural exploitation in exchanges, had shown the degree to which indigenous peoples are particularly disadvantaged in genomic research. In concluding that undue inducement and coercion, as well as other forms of exploitation of vulnerable participants, were primarily to be prevented by effectively functioning RECs, it became clear that indigenous peoples are not only exposed to a distinct type of collective harm, but in addition that the assurance of non-maleficence provided by effective RECs is lacking.

Not only do indigenous peoples form a vital resource for genomic research, but also their unique cosmologies and worldviews contribute towards a form of vulnerability that demands a careful ethical response. From the analyses of vulnerability above, indigenous peoples could be said to be vulnerable internally, in that their world views lead to particular insecurities *vis a vis* the more dominant Western countries, but they are also consent-vulnerable and risk-vulnerable. Some might add justice-vulnerable, referring to their well-documented failures to protect their own financial, cultural or identity-related interests from being harmed. In addition to the usual collective risks of harm associated with genetic research, the potential identity-and privacy-related harms of discrimination and stigmatisation faced by indigenous peoples was shown to be exacerbated in many instances by histories of domination by Western or developed world countries. This history has in many cases demonstrably contributed towards a pervasive distrust of the power and influence of Western researchers.

Analysis of the various guidelines that have emerged in past decades governing research engagement with indigenous communities has enabled the identification of certain core proposals, which if followed by researchers, would address the most serious concerns. The first is the requirement for respectful engagement, which emphasises the need for mutual respect and partnership between researchers and community as equals. The second proposal stresses the need for the inclusion of indigenous peoples in RECs that affect their communities, as well as for the creation of exclusively indigenous RECs. The third proposal is a requirement for institution of community review processes designed to cover all aspects of engagement, permission and consent. These processes bring about the respectful involvement of the indigenous community in key aspects of the consultations, ensuring collaborative and legitimate research. Where benefits are appropriate, these will form part of the research program, and community review would obviate concerns such as undue influence or coercion in the collective consent. The entire process could be regarded as an assurance of *authenticity,* as well as of procedural and substantive justice in research with indigenous communities.

The question as to whether concerns of indigenous peoples are adequately met by the current research practices and guidelines cannot be answered in the affirmative. The guidelines described above and this book' proposals are useful steps in the right direction, but effective application will be required in order to ensure ethical research. For example, mere representation on a REC by one or more indigenous representatives is no guarantee that the legitimate concerns of individual 'nested' communities will be met, and proposals for exclusive indigenous RECs, expensive to institute, have generally not found wide acceptance or budgetary support in practice. A further less than positive outlook on the issue demonstrated by the San case is the fact that the institutional assurance provided by RECs, even where they do exist, is in practice far from adequate for indigenous peoples.

Further legitimate apprehensions about reliance on RECs in the developing world stem from the lack of government support, funding, legislation and applicable guidelines in many countries, as well as other internal components of poor capacity. All of these problems contribute towards the continued vulnerability of indigenous peoples, as well as their potential mistrust of genomic research. The need to address these inadequacies should become an explicit goal in genomic research, and should encourage stakeholders to strive for enhanced ethical assurance. Increased awareness of the legitimate apprehensions of indigenous peoples, plus institution of the proposals suggested above, would contribute significantly towards procedural as well as substantive justice in this arena of genomic research. Representatives of research organisations as well as all members of RECs entrusted with such issues should therefore be reminded of the particular responsibilities activated when vulnerable communities are involved. Indigenous peoples, as an important category of such vulnerable communities, should be reliably assured of the ethical principles of justice, autonomy, beneficence and non-maleficence in genomic research.

References

Abadian, S. (1999). *From wasteland to homeland: trauma and the renewal of indigenous peoples and their communities*. Unpublished Ph.D. dissertation. Cambridge, MA: Harvard University.

ACUNS. (1982). Association of Canadian Universities for Northern Studies. *Ethical principles for the conduct of research in the north*. Ottawa: ACUNS. http://acuns.ca/website/wp-content/uploads/2010/09/EthicsEnglishmarch2003.pdf.

American Indian Law Centre. (1994). Model tribal research code. *With materials for tribal regulation and checklist for indian health boards*. Albuquerque, New Mexico.

Annas, G. (1993). Privacy rules for DNA Databanks: Protecting coded 'future diaries'. *The Journal of American Medical Association, 270*(19), 2346–2350.

Annas, G., Glantz, L. H., & Roche, P. A. (1995). Drafting the genetic privacy act: Science policy and practical consideration. *Journal of Law, Medicine and Ethics, 23*(4), 360–366.

AIATSIS. (2010). Research ethics guidelines. Institute of aboriginal and torres straits islanders studies. http://www.aiatsis.gov.au/research/ethics/index.html.

Baker, Marie Annharte. (1990). *Cheeky moon. Gatherings: The En'owkin journal of First North American peoples 1 (Fall 1990) 38. History of Indian-white relations*. Washington, DC: Smithsonian Institution.

Bannister, K. (2005). Building healthy communities: The role of community-based research. In *Proceedings from the University of Victoria Forum held on*. Retrieved April 23, 2005, from www.polisproject.org/PDFs/CBRForumproceedings.pdf.

Bannister, K., & Hardison, P. (2006). *Mobilizing traditional knowledge and expertise for decision-making on biodiversity*. Case study paper. Consultative Process. www.polisproject.org/PDFs/BannisterHardison.

Begbie-Clench, B. (2010). *Letter to vanessa hayes*. Email to Vanessa Hayes. Dates 17th May 2000.

Benatar, S. (2000). Avoiding exploitation in clinical research. *Cambridge Quarterly of Healthcare Ethics, 9*, 562–565.

Benatar, S. (2004). *International research ethics: Towards a new paradigm*. http://www.tbethics.org/conf17.htm.

Brave Heart, M. Y. H., & De Bruyn, L. (1998). The American holocaust: Historical unresolved grief among native American Indians. *National Center for American Indian and Alaska Native Mental Health Research Journal, 8*, 56–78.

Brody, H. (2001). *The other side of Eden: Hunter-gatherers, farmers, and the shaping of the world.* Canada: North Point Press.

Callaway, E. (2011). Aboriginal genome analysis comes to grips with ethics. *Nature, 477*, 522–523.

Castanello, M. B. (2004). Ethics of aboriginal research. *Journal of Aboriginal Health, 1*, 98–114.

Caulfield, T. A. (2003). Genetic technologies, health care policy and the patent bargain. *Clinical Genetics, 63*(1), 15–18.

Cavalli-Sforza, L., Wilson, A. C., Cantor, C. R., Cook-Deegan, R. M., & King, M. C. (1991). Call for a worldwide survey of human genetic diversity: A vanishing opportunity for the human genome project. *Genomics, 11*, 490.

Chennells, R., Haraseb, V., & Ngakaeaja, M. (2009). Speaking for the San: Challenges for representative institutions. In R. Wynberg, D. Schroeder, & R. Chennells (Eds.), *Indigenous peoples, consent and benefit sharing: Lessons from the San Hoodia Case* (pp. 165–189). Springer: London.

Carlyle, S. W. (2005). Discerning the origins of the Anasazi with mtDNA haplogroups. In D. M. Reed (Ed.), *Biomolecular archaeology: Genetic approaches to the past* (pp. 93–127). Carbondale: Southern Illinois University.

Cobo, M. (2010). *Study on the problem of discrimination against indigenous population.* http://undesadspd.org/IndigenousPeoples/LibraryDocuments/MartínezCoboStudy.aspx.

COHRED. (2013). *Council on Health Research for Development.* Pacific Island publications in the reproductive health literature 2000–2011. www.COHRED.org/publications.

Davis, D. S. (2000). Groups, communities and contested identities in genetic research. *The Hastings Centre Report., 30*(6), 38–45.

De Vries, J., Bull, S., Doumbo, O., Ibrahim, M., Mercereau-Puijalon, O., Kwiatkowski, D., & Parker, M. (2012). Ethical issues in human genomics research in developing countries. *BMC Medical Ethics, 12*(5), 1–10.

Diallo, D. A., Doumba, D. K., Plowe, C. V., Wellems, T. E., Emanuel, E. J., & Hurst, S. A. (2005). Community permission for medical research in developing countries. *HIV/AIDS Clinical Infections Diseases, 41*, 255–259.

Dodson, M., & Williamson, R. (1999). Indigenous peoples and the morality of the human genome diversity project. *Journal of Medical Ethics, 25*, 204–208.

Drabiak-Syed, K. (2010). Lessons from havasupai tribe v arizona state university board of regents; Recognising group, cultural and dignitary harms as legitimate risks warrinting integration into research practice. *Journal of Health and Biomedical Law, VI*, 175–225.

Foster, M. W., Sharp, R. R., Freeman., W, L, Chino, M, Bernsten, D., & Carter. R. H. (1999). The role of community review in evaluating the risks of human genetic variation research. *American Journal of Human Genetics, 64*, 1719–1727.

Freeman, W. L. (1998). The role of community in research with stored tissue samples. In R. F. Weir (Ed.), *Stored tissue samples: Ethical, legal and public policy implications* (pp. 267–294). Iowa City: University of Iowa Press.

Fujimura, J., & Rajagopalan, R. (2011). Different differences: The use of 'genetic ancestry' versus race in biomedical human genetic research. *Social Studies of Science, 41*, 5–30. http://sss.sagepub.com/content/41/1/5.

Ginsburg, D. H. (1999). Genetics and privacy. *Texas Review of Law and Politics, 4*, 17–23.

Godard, B., Marshall, J., Laberge, C., & Knoppers, B. M. (2004). Strategies for consulting with the community: The cases of four large-scale genetic databases. Science and Engineering Ethics, *10*, 457–477.

Greely, H. T. (1997). The control of G3netic research: Involving the "Groups between". *Houston Law Review, 33*, 1397–1430.

Greely, H. T. (2000). Iceland's plan for genomics research: facts and implications. *Jurimetrics, 40*(2), 153–191.

Greely, H. T. (2001). Informed consent and other ethical issues in human population genetics. *Annual Review of Genetics, 35*, 785–800.

Grounds, R. (1996). The Yuchi community and the human genome diversity project: Historic and contemporary ironies. *Genes People and Property, Cultural Survival*, 20(2) summer 1996. http://www.culturalsurvival.org/publications/cultural-survival-quarterly/united-states/yuchi-community-and-human-genome-diversity-project.

Harry, D., Kanehe, L. (2006). Asserting tribal sovereignty over cultural property. *Seattle Journal of Social Justice, Fall/Winter*, 27–66.

Harry, D. (2009). Indigenous peoples and gene disputes. *Chicago-Kent law review, 84*, 179–193.

Hart, S., & Sobraskne, K. A. (2003). *Investigative report concerning the medical genetics project at Havasupai.* December 23. Investigative report available at the Arizona State University, Ross-Baker Law Library, Reserves. http://www.tandfonline.com/doi/abs/10.1080/21507716.2013.770104?journalCode=uabr20.

Hayes, V. (2010). Email correspondence. *Ben Begbie-Clench.* Sent to Ben Begbie-Clench on May 11th 2010.

HGDP. (1997). Human genome diversity project. *Model ethical protocol for collecting DNA samples.* http://www.stanford.edu/group/morrinst/hgdp/protocol.html.

Hook, G. R. (2009). Warrior genes and the disease of being Maori. *MAI Review*, 2, 222–243. http://www.review.mai.ac.nz/index/php/mr/article/viewfile/222/243 Accessed July 2013.

ICC. (1992). *Innuit Circumpolar Conference*, Principles and elements on northern scientific research. In Principles and Elements for a Comprehensive Arctic Policy. Montreal. Centre for Northern Studies and Research: 1992, 142–6.

IPHRC. (2005). *Indigenous peoples' health research centre. Indigenous research ethics*, Report by Willie Ermine to the Institute of Aboriginal People's Health and the Canadian Institutes of Health: March 2005: Saskatoon, SK, Canada.

Isselmuiden, C., & Faden, R. (1992). Research and informed consent in Africa—Another look. *The New England Journal of Medicine, 326*(12), 830–833.

Juengst, J. (1998). Groups as Gatekeepers to genomic research: conceptually confusing, morally hazardous, and practically useless. *Kennedy Institute of Ethics Journal, 8*, 183–200.

Karlawish, J. (1998). Building Right Reason in the Republic of Science: Why we must include community values in the design and conduct of clinical research. *Penn Bioethics, 5*(2), 1.

Karlsen, J. R., Solbakk, J. H., & Holm, S. (2011). Ethical endgames: Broad consent for narrow interests; open consent for closed minds. *Cambridge Quarterly of Healthcare Ethics, 20*(4), 572–583.

Kemp, B. M. (2007). Genetic Analysis of early Holocene skeletal remains from Alaska and its implications for the settlement of the Americas. *American Journal of Physical Anthropology, 132*, 605–621.

Kowal, E. (2012). Genetic research in indigenous health: Significant progress, substantial challenges. *Medical Journal of Australia. 197*(1), 2012.

Lee, R., & Chambers, G. (2007). Monoamine oxidase, addiction and the 'warrior' gene hypobook. *New Zealand Medical Journal, 120*, 2441.

Laird, S. (2002). *Biodiversity and traditional knowledge.* London: Earthscan Publications Limited.

Mander, J., & Tauli-Corpuz, V. (Eds.). (2005). *Paradigm wards: indigenous peoples' resistance to economic globalization. International forum on globalization committee on indigenous peoples*. San Francisco: Sierra Club Books.

McCulloch, A., & Wilkins, D. (1995). "Constructing" nations within states: the quest for federal recognition by the Catawba and Lumbee tribes. *American Indian Quarterly, 19*(3), 361–388.

McGregor, J. L. (2007). Population genomics and research ethics with socially identifiable groups. *Journal of Law, Medicine and Ethics, 35*, 356–370.

McMillan, J. R., & Conlon, C. (2004). The ethics of research related to health care in developing countries. *Journal of Medical Ethics, 30*, 204–206.

Mead, A., & Ratuva, S. (2007). *Pacific genes and life patents: Pacific indigenous experiences and analysis of the commodification and ownership of life*. Wellington: Call of the Earth Llamado de la Tierra and the United Nations University Institute of Advanced Studies. http://calloftheearth.files.wordpress.com/2009/07/coe-publication-final.pdf.

Murray, T. H. (1997). Genetic exceptionalism and 'future diaries.' Is genetic information different from other medical information? In M. A. Rothstein (Ed.) *Genetic secrets. Protecting privacy and confidentiality in the genetic era* (pp. 60–73). New Haven: Yale University Press.

NBAC. (1999). National advisory bioethics commission. *Research involving human biological materials: Ethical issues and policy guidance.* https://bioethicsarchive.georgetown.edu/nbac/pubs.html.

Ngakaeaja, M. (2011). *Letter addressed to editor of Nature*, Deputy Director, Working Group of Indigenous Peoples of Southern Africa. (WIMSA). Retrieved February 18, 2011, from www.wimsa.org.

NHMRC. (1991). *National health and medical research council of Australia; guidelines on ethical matters in aboriginal and Torres strait islander research.* Canberra: Publication no E 13. http://www.nhmrc.gov.au/guidelines/publications/e52.

NHMRC. (2003). *National health and medical research council of Australia; guidelines on ethical matters in aboriginal and Torres strait islander research.* Camberra: Publication no E 13. https://www.nhmrc.gov.au/_files_nhmrc/publications/attachments/e52.pdf

Ouwe-Missi-Oukem-Boyer, O., Munung, N. S., Ntoumi, F., & Tangwa, G. (2013). Capacity building in health research ethics in Central Africa: Key players, current situation and recommendations. *Bioethics forum 6*(1), 1–11.

Olsen, E. G. (1967). *Indian blood.* Parson: McClain Printing Company.

Pillai, P. (1999). Feminism and the problem of epistemic displacement: Reconstructing Indigenous Theories. In A. Kibbey et al. (Eds.) *On your left: The new historical materialism* (pp. 206–247). New York: New York University Press.

Pogge, T. (1997). *Group rights and ethnicity in ethnicity and group rights: Nomos XXXIX 1.* In I. Shapiro, & W. Kymlicka (Eds.), (pp 187–221), New York: New York Universityy Press.

Pullman, D., & Nicholas, G. (2011). Intellectual property and the ethical/ legal status of human DNA: The (ir)relevance of context. *Inuit Studies, 35*(1 & 2), 143–164.

RCAP. (1992). *Royal commission on aboriginal peoples.* Report of the royal commission on aboriginal peoples. Ottawa, ON: Canada Communication Group.

Reardon, J., & Tallbear, K. (2012). Your DNA is our history: genomics, anthropology and the construction of whiteness as property. *Current Anthropology, 53*(supplement 5), 233–245.

Reilly, P. R. (1998). Rethinking risks to human subjects in genetic research. *American Journal of Human Genetics, 63*, 682–685.

Royal Commission on Aboriginal Peoples. (1996). *Report of the royal commission on aboriginal peoples.* Ottawa, ON: Canada Communication Group.

Sharp, R., & Foster, M. W. (2000). Involving study populations in the review of genetic research. *Journal of Law, Medicine and Ethics., 28*, 41–51.

Sharp, R., & Foster, M. W. (2002). Community involvement in the ethical review of genetic research: Lessons from American Indian and Alaska Native populations. *Environmental Health Perspectives, 110*(supplement 2), 145–148.

Schuster, R., Schuster, S. C., Miller, W., et al. (2010). Complete Khoisan and bantu genomes from Southern Africa. Complete Khoisan and bantu genomes from Southern Africa. *Nature, 18*, 463, 943–7. http://www.nature.com/nature/journal/v463/n7283/full/nature08795.html.

Soskolne, C. (1997). Ethical, social and legal issues surrounding studies of susceptible populations and individuals. *Environmental Health Perspectives, 105*(supplement 4), 837–841.

SPEAR. (2008). *Social policy evaluation and research committee (2008) good practice guidelines.* http://www.spear.govt.nz/good-practice/background/index.html.

Spaak, T. (2006). Genetic discrimination. *Minnesota Journal of Law, Science Technology, 7*, 639–655.

Struewing, J., Hange, P., Wacholder, S., Baker, S., Berlin, M., McAdams, M., et al. (1997). The risk of cancer associated with specific mutations of BRCA1 and BRCA2 among ashkenazi jews. *New England Journal of Medicine, 336*, 1401–1408.

Suter, S. (2001). The allure and peril of genetic exceptionalism: Do we need special genetics legislation? *Washington University law Quarterly, 79*(3), 669.

TCPS. (1998). Tri-council policy statement. *By the Natural Sciences and Engineering Research Council, (NSERC) Social Sciences and Humanities research Council, (SSHRC) and Canadian Institutes of Health Research (CIHR).* http://www.nserc.ca/programs/ethics/english.htm.

Tyler, T. R. (1987). Procedural justice research. *Social Justice Research, 1*(1), 41–66.

Tsosie, R. (2007). Cultural challenges to biotechnology: Native American genetic resources and the concept of cultural harm. *Journal of Law, Medicine and Ethics, 35*(3), 396–411.

Tsosie, R., & McGregor, J. L. (2007). Genome justice: Genetics and group rights. *Journal of Law, Medicine and Ethics., 35*(3), 352–363.

UN. (1966). *International covenant on economic, social and cultural rights.* http://www.ohchr.org/EN/ProfessionalInterest/Pages/CESCR.aspx.

UNDRIP. (2007). *United Nations declaration on the rights of indigenous peoples.* www.un.org/esa/socdev/unpfii/documents/DRIPS.

UNESCO. (1990). *Meeting of experts on future study of the rights of peoples.* Paris: UNESCO.

UNESCO. (1995). International bioethics committee. *Bioethics and human population genetics research.* Paris: UNESCO. Retrieved November 15, 1995. www.portal.unesco.org/ulis/index.html.

Waitangi Tribunal. (2011). *Ko Aotearoa Tenei: A report into claims concerning New Zealand law and policy affecting maori culture and identity.* Report 262. http://www.waitangi-tribunal.govt.nz/reports/summary.asp?reportid={BF981901=5B55.

Wang, Y. (2011). *Group protection in human population genetic research in developing countries: the People's Republic of China as an example.* Unpublished Ph.D. book. University of Glasgow. http://theses.gla.ac.uk/3005/1/2011WangYuephd.pdf.

Weijer, C. (1999). Protecting communities in research: Philosophical and pragmatic challenges. *Cambridge Quarterly for Healthcare Ethics, 8*, 503–513.

Weijer, C. (2000b). Benefit sharing and other protections for communities in genetic research. *Clinical Genetics, 58*(5), 367–368.

Weijer, C., Golds, G., & Emanuel, E. J. (1999). Protecting communities in research: current guidelines and limits of extrapolation. *Nature Genetics, 23*, 275–280.

Weijer, C., & Miller, P. B. (2004). Protecting communities in pharmacogenetic and pharmacogenomic research. *The Pharmacogenomics Journal, 4*, 9–16.

Wells, S. (2007). *Deep ancestry: Inside the genographic project.* Washington, DC: National Geographic Books.

Wesley-Esquimaux, C. C. & Smolewski, O. (2004). Historic trauma and aboriginal healing. Ottawa, ON: Aboriginal Healing Foundation. http://www.ahf.ca/downloads/historic-trauma.pdf.

Wynberg, R., & Chennells, R. (2009). Green diamonds of the south: An overview of the San-Hoodia case. In R. Wynberg, D. Schroeder, & R. Chennells (Eds.) *Indigenous peoples, consent and benefit sharing: Lessons from the San Hoodia case* (pp. 89–124). London: Springer.

Chapter 9
Closing Chapter

> *The lion may well lie down with the lamb, but the lamb will not get a good night's rest.*
>
> (Attributed to Woody Allen)

Abstract The closing chapter reframes the initial questions posed in this book in a fresh manner, aimed at providing a clear statement of the main arguments and the final positions reached. An imaginary conversation with a San leader leads to the extraction of three overall questions which should be answered in the affirmative when engaging in genetic research with vulnerable peoples. Firstly, is there fair treatment? Secondly, is there a fair deal? And thirdly, is there assurance that the research will cause no harm? After discussing each of these, a diagram depicts the three overall questions, against the main ethical principles applicable to genetic research, in relation to the two primary forms of enquiry, namely into the outcome of the research, and into the procedure followed. The book concludes with a decisive reply to the statement on exploitation posed by the imaginary critic in the first chapter.

Keywords Vulnerable populations · Ethically fair genetic research · Exploitation in genetic research

The wry quote above might appear rather informal for a serious academic book, but it has been chosen to reflect the latent power asymmetry that so often accompanies abuse of the weak in practice. After decades as a human rights lawyer seeking to correct institutional bullying or abuses of power by strong over vulnerable parties, latterly assisting indigenous peoples on benefit sharing under the CBD, I became increasingly disillusioned with the limitations of the law in correcting such wrongs. So many transactions take place in rural locations, far from watching eyes or sound advice, and under circumstances conducive for manifestly unequal engagements. David versus Goliath stories, such as the Havasupai case against the University of Arizona, are consequently few and far between. Whilst acting as a lawyer, I became aware of similar problems experienced by indigenous peoples targeted for genomic

R.S. Chennells, *Equitable Access to Human Biological Resources in Developing Countries*, DOI 10.1007/978-3-319-19725-8_9

Table 9.1 Research questions, main ethical principles, two forms of enquiry

Three research questions	Main ethical principles	Outcomes enquiry	Procedural enquiry
1. Is benefit sharing necessary?	A Fair Deal: Justice (do not exploit)	*Apply*: Justice in exchange and Mayer's 2nd test for equitable exchange. *The test*: Benefits should be a proportionate requital to the sample donation	
2. Does benefit sharing lead to undue inducement and coercion?	Fair treatment: Protection of vulnerable populations (do not coerce)		*Apply*: The Law of Equity. *The test*: Did the method of dealing impair consent in terms of: 1. Unconscionable dealing? 2. Duress/coercion? 3. Undue Influence?
3. Are the concerns of indigenous peoples adequately met by current practices and guidelines?	Primum non nocere (do no harm)	*Apply*: Mayer's 3rd test for authentic research. *The test*: REC must provide assurance of do no harm principle	
	Respect for communities		*Apply*: Consult on community values and choices. *The test*: REC must ensure indigenous representation or indigenous REC.

This table is not to be understood as an Ethical Matrix as suggested by Ben Mepham based on Principlism. Its purpose is to capture the conclusion in one diagrammatic illustration research. Effective policies and ethical practices were what was required, rather than the application of formal law. I was introduced to bioethics as a burgeoning field, and decided to approach the issues addressed below in the context of applied philosophy and public ethics rather than in the field of law (Table 9.1).

In continuation of this autobiographical beginning, the essential work of this book will be introduced by means of an imaginary conversation between an indigenous leader Andriemario and his legal advisor Rotman, based vaguely upon living persons, in which they deliberate about a meeting that they are about to hold with a genomic researcher. The structure for concluding this book will emerge from the following discussion.

Andriemario:	So Rotman who is this person coming to meet us and what does she want?
Rotman:	She is called doctor Simla. She is a famous genomics researcher I believe, and wants to know if the San will participate in a genomic research program. She will explain it all to you I am sure.
Andriemario:	OK so I suppose she wants to take the blood of the San people far away and publish a lot of lies about us and become even more famous than ever and make a lot of money as well, just like all the others have done?" Maybe she'll offer us cigarettes?
Rotman:	Are you being serious? Anyway I am here to assist you make sure that does not happen. Now I just need to know from you Andriemario, what are your main concerns about the whole genomic research thing? Just so I can prepare myself.
Andriemario:	Ag Rotman you know us well. We are not trying to be difficult. But we just want to be treated properly, the deal must be fair, and the whole research thing must not cause any harm for any of our people.
Rotman:	OK so what exactly do you mean by those things? What is being 'treated properly'?
Andriemario:	It's simple man. We just need to be treated with respect. They must not hurry us or force us or lie to us. They must give us all the information so we can understand and make a good decision on our own. They must listen to us carefully and then respect our views on what is important. Not like those Schuster researchers did with the San in Namibia.
Rotman:	OK I've got that. Now what do you mean by a 'fair deal'?
Andriemario:	This all depends. It must just be fair and balanced. If they are making money, they should offer a share to us. If they have some money, they should assist our community in some way for our contribution. If they have no money, and the research is a good thing for our people or for other people, we might still participate.
Rotman:	I've got you Andriemario. Now what do you mean by the 'no harm' part?

Andriemario: Eish! Sometimes you ask silly questions, Rotman. How can we know until we hear more from this clever Dr Simla what she intends to do with our blood. If she is going to publish information, then we need to know what information they are looking for, and where they will publish it, so we can make sure that bad things are not said about us. We are sensitive about some things, and just don't want bad things to happen for our people.

One can imagine variations of these three concerns expressed by the indigenous leader depicted above being replicated across different continents whenever genomic research encounters are envisaged. This book has endeavoured to approach the research topic within the parameters of the research ethics domain constituted by the applicable laws, guidelines and principles, the interpretations of which are influenced in no small measure by prominent commentators in the literature. The overall questions set out at the start of this book were aimed to address precisely the three concerns expressed by Andriemario above, in a different order, and can be reformulated as being firstly a plea for fair treatment, secondly for a fair deal, and thirdly for assurance of research that gives rise to no harm.

9.1 Fair Treatment

Fair treatment implies a fair process. The research ethics guidelines, created to ensure ethical research on humans, warn against undue inducement and coercion as being two of the ways that individuals can be exploited in the process prior to giving informed consent. Thomas Pogge's schematic conception of justice was drawn upon to show that procedure is a core component of every dimension at which justice plays out, including the particular form known as commutative justice, which includes contract law. Equity law infused contract law with very clear criteria for ascertaining procedural justice, ensuring that a vulnerable party to a contract could be protected if she were treated unfairly. Unfair treatment in equity law was interpreted as some form of inappropriate dealing in which the vulnerable party was either wrongly manipulated, coerced, or influenced during the process leading up to the provision of legal consent. The holistic enquiry required by equity law showed that no aspect of the transaction, neither vulnerability nor extreme need nor other imbalances of power could stand alone from other factors, and needed to be weighed up in conjunction with relationships, motives and actions. In explaining the elements of unconscionable dealing, duress and undue influence, the precise manner of the dealing was in each case central to the enquiry into fair treatment, and the actual outcome was shown to be of no more than peripheral import.

The ethos of fair procedures embodied in the requirements of equity law resonated strongly with recommendations for more equitable treatment of indigenous peoples referred to in the eighth chapter. Respect for communities, equal partnership, participative processes, and community review were all components of such fair processes. RECs overseeing research involving indigenous peoples were crucial for the improvement of such processes, and proposals for representation of communities on the important bodies were described. These proposals, now captured in the growing body of guidelines that aim to ensure ethical research involving indigenous peoples, resonate with the deceptively simple request expressed by the imaginary Andriemario above, namely that they merely wish to be treated 'properly.'

To conclude discussion of this first procedural concern, the dangers of undue influence and coercion that were so prominent in the clinical domain were shown to be relatively minor with regard to genomic research. The ethical principle of respect for communities needs to be fleshed out and given practical effect, ensuring that indigenous peoples are engaged with sensitively and as equal partners in genomic research.

9.2 A Fair Deal

Seeking acceptable criteria on a fair distribution of benefits and burdens has been a preoccupation of philosophers and jurists for centuries, and this book did not attempt to solve the issue. How to deal with unequal bargaining power and vulnerable parties was one of Aristotle's preoccupations whilst writing on justice, and remains a central concern of theorists on the nature of exploitation. The balance between freedom of contract and autonomy on the one hand and restrictive rules and laws on the other in the public interest remains a complex aspiration sought after by the applicable guidelines. Two particular benchmarks were selected in order to approach the question as to what would be a fair deal for an indigenous community engaged in research, namely commutative justice, and Mayer's second class of exploitation.

Aristotle's concept of commutative justice, which governs voluntary transactions between two parties, was consistently referred to as one of the main reasons for the negotiation of the CBD, as well as a motivation for the institution of benefit sharing in relation to research involving genetic resources. The history of and reasons leading to the CBD were treated to an in-depth analysis, during which it became apparent that bioprospecting of human and non-human DNA are similar and parallel in every way. It was shown conclusively, I suggest, that the high research value of both forms of DNA in the hands of receivers justified both the imposition of benefit sharing as a legal requirement of the CBD, as well as the subsequent extension of benefit sharing in relation to human genetic resources.

After examining various theories relating to exploitation, Mayer's exposition of the concept based upon fairness was drawn upon to provide a method of assessing fairness in exchange between two parties. Mayer's second class of exploitation, which takes place when one party benefits unfairly at the expense of the other, resonates in every way with Aristotle's commutative justice, both of which require some form of requital in order to correct an unequal transaction with a weaker party. Genomic research involving humans was shown to be motivated by exactly the same drive to patent and commercialise as that which fuels bioprospecting of other genetic resources. Despite the important contributions made by the open access movement, it was evident that registration of patents over all forms of DNA-related inventions plays a ubiquitous and prominent role in the research process. Mayer's test of fairness enabled the point to be emphasised that even if research communities do benefit in some minor way, eventually, the real question was whether they received a commensurate or appropriate level of benefits in view of the research value of the samples.

The assessment of a fair deal required under these tests, or otherwise stated the attempt to ensure beneficence in a transaction, requires a substantive or outcomes-based test. Aristotle envisaged a process in commutative justice where the respective benefits were the focus of the enquiry, rather than the motives, dealings or relationships of the parties. As was described above, finding the right criteria to establish fairness of an outcome is no simple matter, but the calculation nevertheless needs to be made. Wages paid to sweatshop employees in developing world countries can be judged as exploitative, even if the values and benchmarks for fairness are set elsewhere. The Tristan da Cunha islanders received nothing from profits made by Sequana Therapeutics, which enabled commentators to formulate a prima facie view that sufficient requital had not been made.

The words of the imaginary Andriemario in the above conversation saying that what would be a fair deal would *depend* on the circumstances was thus an indication that a common-sense or practical process was able to establish the most appropriate criteria for fairness. All in all, with the assistance of a fair procedure enabling exchange of information, the tests for justice in exchange proposed by Aristotle and Mayer are suggested to be clear enough to be applied in practice, and support the proposals for benefit sharing in genomic research.

A final hurdle to be crossed was the fact that various objections have been made against the sharing of benefits with research participants. These arguments, namely the common heritage, the altruism and the no value added arguments were all dealt with extensively, and can be safely regarded as being disposed of as impediments to benefit sharing. As a final approach to what is a fair deal, the words of Andriemario above are reassuringly practical. The nature and extent of benefits should depend entirely upon the circumstances.

9.3 Research that Entails no Harm

This third requirement brings together elements of both procedural and substantive justice. The procedural aspect is covered by the recommendations for ethical engagement with indigenous peoples described in the eighth chapter, including the community review process and representation of indigenous peoples on the appropriate REC. The substantive aspect on the other hand is covered by the REC assessment that the research should benefit the recipients authentically, in other words it should pass the test of non-maleficence. The five broad themes of engagement synbooked from independent indigenous documents by Charles Weijer provide ample guidance in the procedural regard, and are designed to establish trust and ascertain any particular cultural sensitivities relevant to the research. During such an engagement, all information relating to the research is to be exchanged and fully understood, ensuring that the outcome reached by such a legitimate procedure is authentic and approved of.

Employment of sound participative and community review processes will not only lead to a relationship of trust with the indigenous community, but are designed to establish the risks of harm associated with the proposed research, namely the desired outcome of substantive justice. The leader Andriemario stated above, 'we are sensitive about some things,' a reminder that a REC's assurance of non-maleficence can only be provided with full knowledge of the cultural context. Whilst the approval of an appropriate research protocol remains the ultimate responsibility of the REC, issues of representation, lack of capacity, lack of funding support and other factors contribute to the less than satisfactory state of the RECs in the developing world. According to the research ethics guidelines a REC should only approve research if it is authentic in every way, namely that it does no harm, and that it complies with all the requirements for ethical research on the particular participant community. Nothing short of proper procedures and authentic outcomes by RECs are able to guarantee that indigenous peoples are protected in genomic research. The ethical guidelines, principles and protocols are largely in place, whilst legitimate concerns exist with regard to the lack of capacity of these important institutions to carry out their important roles.

9.4 Closing Remarks

The three research questions, juxtaposed with Andriemario's three questions as discussed above, are schematically depicted alongside the procedural and substantive enquiries brought to bear. The diagram above shows the essential criteria to be applied at each stage, and also the research ethics principles most applicable in each enquiry.

Finally and in response to the words of the imaginary critic recorded in the introductory chapter, this book is now in a position to reply firmly as follows:

> Exploitation remains a potential problem in genomic research, for indigenous peoples as well as other communities in the developing world. Benefit sharing is essential in order to compensate for the value of resources imparted, and to bring about justice in exchange. The objections to benefit sharing (the common heritage of humankind argument, the altruism argument, and the 'no value added' arguments) were addressed and found to lack coherence. As a second point is must be added that the requirement of benefit sharing in genomic research is unlikely to lead to undue inducement or coercion of research participants, a danger far more applicable to the realm of clinical research. Finally, the legitimate concerns of indigenous peoples are not adequately met either by current guidelines or by the RECs, which are currently lacking in capacity and therefore unable to ensure non-maleficence. Adoption of the proposals made in this book will alleviate these concerns.

List of Legal Cases

Allcard v Skinner (1887) AC 36 Ch D 145 [Legal Case]
Baden v Societe Generale pour Favoriser le Developpement du Commerce et l'Industrie en France (1982, 1992) All ER 161
Bank of Montreal v Stuart (1911) AC 120
Blomley v Ryan (1956) 99 CLR 362 HCA 81
Bray v Ford (1986) AC 44
Bridgewater v Leahy (1998) 194 CLR 457
Commercial Bank of Australia Limited v Amadio (1983) 151 CLR 447
Diamond v Chakrabarty (1980) 447 US 303 (1980)
Diprose v Louth (no 2) (1990) 54 SASR 450
Earl of Chesterfield v Janssen (1751) 2 Ves Sen 125:28 ER 82
Harry v Kreutziger (1978) 95 DLR (3rd) 231
Johnson v Butress (1936) 56 CLR 113
Lloyds Bank vs Bundy (1975) EWCA Civ 8
McPherson v Watt (1877) LR 3 App Cas 254
Patel vs Grobelaar (1974) SA 532 A
Peanut Marketing Board v Cuda (1984) 79 FLR 368
Royal Bank of Scotland plc v Etridge (no 2) (2001) 3 WLR 1021
Tito v Waddell (1977) no 2: Chancery 106
Treitel (1999) law of Contract (ref please)
Wright v Carter (1903) 1 Ch 27

R.S. Chennells, *Equitable Access to Human Biological Resources in Developing Countries*, DOI 10.1007/978-3-319-19725-8

MIX
Papier aus verantwortungsvollen Quellen
Paper from responsible sources
FSC® C105338

If you have any concerns about our products,
you can contact us on
ProductSafety@springernature.com

In case Publisher is established outside the EU,
the EU authorized representative is:
Springer Nature Customer Service Center GmbH
Europaplatz 3, 69115 Heidelberg, Germany

Printed by Libri Plureos GmbH
in Hamburg, Germany